V.F. Krotov, V.Z. Bukreev, and V.I. Gurman

NEW VARIATIONAL METHODS IN FLIGHT DYNAMICS

(Novye metody variatsionnogo ischisleniya v dinamike poleta)

Izdatel'stvo "Mashinostroenie"
Moskva 1969

Translated from Russian

Israel Program for Scientific Translations
Jerusalem 1971

TT 70-50186
NASA TT F-657

Published Pursuant to an Agreement with
THE NATIONAL AERONAUTICS AND SPACE ADMINISTRATION
and
THE NATIONAL SCIENCE FOUNDATION, WASHINGTON, D. C.

Translated and edited by IPST Staff

Printed in Jerusalem by Keter Press
Binding: Wiener Bindery Ltd., Jerusalem

Available from the
U.S. DEPARTMENT OF COMMERCE
National Technical Information Service
Springfield, Va. 22151

Table of Contents

PREFACE

The present volume deals with variational methods for the optimization of motion of aircraft and other objects often encountered in modern technology. The development of aeronautics and astronautics progressively focuses ever increasing attention on the determination of optimum flight programs characterized by minimum time, minimum fuel consumption, maximum range, minimum cost, etc. Numerous publications which appeared in the USSR and in the West during the last 20.—25 years deal with various aspects of this optimization problem. The leading contributions were made by I. V. Ostoslavskii, D. E. Okhotsimskii, A. A. Kosmodem'yanskii, and A. Miele. Their work played an important role in the development of controlled flight mechanics; they were also the first to come face to face with the fundamental difficulties that to this date plague us in the solution of modern optimal problems. These difficulties are a result of the complex and varied conditions that must be taken into consideration when formulating and solving the various problems. The following factors fall under this category:

1) various inequality restraints are imposed on the system variables by physical and practical considerations (restraints on altitude and velocity, on the overload, angle of attack, engine thrust, etc.);

2) one is required to determine the absolute minimum, rather than a local (or relative) minimum;

3) the sought optimal control often does not exist in the class of admissible controls in the classical formulation of the problem or, in more general terms, we have to deal with degenerate problems.

The above features are characteristic of all optimal control problems, and are not specific to flight dynamics. The difficulties can be overcome in two ways. The first approach called for modernizing the standard classical methods (transformation of variables, introduction of additional variables, direct analysis of variations). The second approach involved more radical measures, i.e., development of new optimal principles with logic not affected by the above considerations.

It is this second approach that produced the new optimization methods due to R. Bellman (dynamic programming), L. S. Pontryagin (the maximum principle), and the variational methods which constitute the basis of our book.

The monograph presents a systematic review of the authors' results obtained over a number of years. It is divided into six chapters and a Supplement. Chapter I reviews the elements of the theory of variational methods. Unlike the traditional approach, the variational technique also covers the case when the sought optimal control does not exist in the class of admissible controls and the solution must be derived by constructing a minimizing sequence. This is a typical case often encountered in modern practice. The fundamental theorem of the sufficient conditions of the absolute minimum of a functional is then applied to reduce the problem of functional minimization to the problem of maximum of some function R of the problem variables for every fixed value of the argument.

Chapter II describes in detail a number of particular methods. The equations of Pontryagin's maximum principle are derived in § 2.1, and the reader is acquainted with this method from a new angle. These equations are then reinforced with new conditions of a strong local (relative) minimum. Significantly, these conditions are linked with the classical sufficient conditions of variational calculus. Jacobi's famous condition is seen to be equivalent to the requirement of negative-definiteness of the second differential of the function R.

The important concept of the synthesis of optimal control is also dealt with in this chapter. Bellman's equation is derived as a particular case from the conditions of the fundamental theorem, and the different forms of solution of the optimal problem — synthesis and program — are briefly illustrated. The complete solution for the synthesis of linear systems is given. The method of approximate synthesis and, primarily, the a priori upper-bound estimate of the solution accuracy are of the greatest practical importance among these topics. New methods of this kind are of great value, since the Bellman equation often proves too complicated for practical application and its exact solution too problematic.

Chapter III considers the structure and the analysis of degenerate and sliding controls. The treatment starts with a description of the construction of a sliding-control minimizing sequence; the construction is first described for a simple example and then in a more general form, depending on the properties of the function R. The methods of Pontryagin and Bellman are then generalized to the case of degenerate solutions. The last sections of the chapter present a new special method particularly suitable for degenerate and sliding controls, the method of multiple maxima. This method is highly effective for flight dynamic problems.

Also of considerable interest is the theoretical application of this method to the investigation of the degenerate second variation of the functional. This approach leads to a natural generalization of all the standard conditions of variational calculus, including Jacobi's global condition, to the case of degenerate solutions.

This method is still in the development stage, but the results summarized in this book are quite sufficient to illustrate its great potential, in particular, for qualitative analysis of various problems with the purpose of deriving "quick" estimates. This is a highly valuable attribute at the design and synthesis stage, when exact solutions are meaningless, insofar as the system parameters have not been fixed.

The next two chapters deal with applications. Chapter IV presents solutions of a number of problems of powered flight dynamics. First to be considered is the elementary problem of the vertical ascent (descent) of a rocket in vacuum. The well-known solution of this problem is derived here as an illustration of the applications of the new methods (here, as in most other applications, the method of multiple maxima is employed). The general problem of rocket dynamics in a homogeneous field in a vacuum is then considered. The application of the new method leads to an original treatment of this problem and yields new simple equations of optimal motion, with coasting integrals playing the role of variables and the direction of thrust and time figuring as controls.

Next to be considered is the so-called classical problem of flight dynamics, e.g., the problem of powered ascent of an aircraft. It was previously solved by different methods by Ostoslavskii, Egorov, and Miele. The application of our new method yields a more accurate result; in the case of several extremals (e.g., the typical situation in supersonic flight) the transition points between the different extremals are naturally determined.

The last section of chapter IV illustrates a detailed construction of an approximate synthesis of optimal aircraft control on the ascent section. The synthesis of optimal control is the most desirable form of solution of the optimal problem, but it is more time- and labor-consuming than the construction of the optimal program. This feature is probably linked with the infrequent use of synthesis solutions in nonlinear systems.

Chapter V investigates optimal coasting controls of winged aircraft subjected to lift. Minimum-time and minimum-heat descent programs are discussed, using various characteristic restraints (on altitude, angle of attack, temperature, overload, etc.). The method of multiple maxima is again very effective in this case, as it yields approximate optimal solutions (with detailed accuracy estimates) in the form of finite expressions, without involving the engineer in a tedious and difficult solution of boundary-value problems. Note that the sought optimal control is found to be a sliding control for the angle of attack over a considerable length of the trajectory; in practice, it can be realized by a relay control with a moderate switching frequency.

Chapter VI presents an analog of the optimum principle for discrete control systems described by finite-difference equations. Although the treatment is purely theoretical, its practical value is self-evident in the light of ever increasing computer applications. Optimal problems using differential equations are converted into finite-difference schemes for computer solution.

The Supplement at the end of the book is fundamentally different from the earlier chapters. It will be of interest to readers who wish to acquaint themselves with the basic problems of variational calculus and with the fundamental ideas that served as a point of departure for this monograph. The Supplement also contains a useful solution of the problem of maximum-range horizontal flight of an aircraft.

The authors acknowledge the great help of Acad. A. M. Letov, who reviewed the manuscript and made a number of valuable comments. The authors are also grateful to the scientific editor of the book, I. V. Ioslovich, for his efforts.

Reader's comments will be most welcome. Please address all correspondence to "Mashinostroenie" Publishing House, Moscow, K-51, Petrovka 24.

INTRODUCTION

Various problems of science and technology often require choosing
the best, or the o p t i m u m, solution among a set of all the possible
alternative solutions. The mathematical formalism of problems of this
kind generally operates with two concepts: the concept of a s e t and the
concept of a f u n c t i o n a l defined over a given set. Examples of sets
are the set of all real numbers between 0 and 1, the set of all possible
directions of the thrust vector of an aircraft engine, the set of all possible
trajectories taking an aircraft to a given altitude, etc.

A f u n c t i o n a l is said to be defined over some set if we know how to
assign a definite number to each element of the set.

Consider some simple examples of functionals.

1. Consider the set of all plane curves. To each curve we assign a
certain number — its length. The length of a curve is thus functional.

2. Consider a point in the (x, y) plane which may move along any path
between two given points A and B so that it has a definite velocity $V(x, y)$
at every point (x, y) of the path. Assigning to each path of the set the time
to traverse that path, we obtain a functional.

3. The maximum surface temperature of a spacecraft during atmospheric reentry is a functional which depends on the trajectory of the
spacecraft.

The general formalism of choosing optimum solutions (according to
certain "quality" criteria) operates with a fixed set M and a functional
$I(v), v \in M$, defined over the elements of this set which characterizes the
quality of the element v, so that if an element v_1 is "better" than an
element v_2 in a certain sense, $I(v_1) < I(v_2)$ and vice versa. The problem
is thus formulated as follows: find an element v of the set M for which the
functional $I(v)$ has its minimum value $I(v) = m$. This problem, however,
is not always solvable. The set being considered does not always contain
an element v satisfying the exact equality $I(v) = m$. For example, the
function $y = ax + b$, $a > 0$, attains its minimum at the point $x = 0$ on the set of
the points x between 0 and 1, provided this set is closed, $0 \leqslant x \leqslant 1$; the
minimum is not attained on any point of the set if it is open, $0 < x < 1$. A
slight modification in the statement of the problem eliminates the difficulty.
The final formulation reads as follows: consider a set M and a functional
$I(v)$ on this set, $v \in M$. Find a sequence of elements of the set M,
$\{v_s\} \subset M$, for which $I(v_s) \to m$ for $s \to \infty$. The resulting sequence is known
as a m i n i m i z i n g   s e q u e n c e. It always exists, in virtue of the
definition of the exact lower bound $m = \inf_M I$. The minimal element $\bar{v}$ may
be treated as one of such minimizing sequences defined in the form $v_s = \bar{v}$
for all s.

In what follows, we will deal with more tangible problems: determination of the optimal control for objects described by ordinary differential equations or by their discrete analogs — the finite-difference equations. As a preliminary step, we have to consider in some detail the concept of a controlled system (or controlled object).

A typical example of a controlled system is provided by an aircraft.

The vector equations of motion of the center of mass of an aircraft in an inertial frame of reference have the form

$$\left.\begin{aligned} m\dot{\vec{V}} &= \vec{F}; \\ \dot{\vec{r}} &= \vec{V}; \\ \dot{m} &= -\beta_f, \end{aligned}\right\} \tag{I.1}$$

where $\vec{r}$, $\vec{V}$ are respectively the radius vector and the velocity vector of the center of mass of the aircraft; m is the mass; β_f is the mass (propellant) consumption per second; F is the resultant vector of all the external forces.

When solving the equations of flight dynamics, the motion of an aircraft is generally considered in the wind system of coordinates /4/:

$$\left.\begin{aligned}
\dot{V} &= \frac{1}{m}[P\cos(\alpha-\varphi)\cos\beta - X\cos\beta + Z\sin\beta - \\
&\quad - G\sin\theta]; \\
\dot{\theta} &= \frac{1}{mV}\{P[\sin(\alpha-\varphi)\cos\gamma_c + \cos(\alpha-\varphi)\sin\beta\sin\gamma_c] - \\
&\quad - X\sin\beta\sin\gamma_c + Y\cos\gamma_c - Z\cos\beta\sin\gamma_c - G\cos\theta\}; \\
\dot{\psi}_c &= \frac{1}{mV\cos\theta}\{P[\sin(\alpha-\varphi)\sin\gamma_c - \cos(\alpha- \\
&\quad -\varphi)\sin\beta\cos\gamma_c] + X\sin\beta\cos\gamma_c + Y\sin\gamma_c + \\
&\quad + Z\cos\beta\cos\gamma_c\}; \\
\dot{h} &= V\sin\theta; \quad \dot{m} = -\beta_f; \quad \dot{x} = V\cos\theta,
\end{aligned}\right\} \tag{I.2}$$

where V is the velocity; h the altitude; x the range on the Earth's surface; θ the angle of inclination of the trajectory to the local horizon; α the angle of attack; ψ_c the angle of yaw; β the angle of side slip; γ_c the angle of bank; φ the angle between the thrust vector and the velocity vector; P engine thrust; G aircraft weight; X, Y, Z the drag, the lift, and the side force, respectively:

$$X = c_x(M, \alpha)\frac{\varrho(h)V^2}{2}S; \quad Y = c_y(M, \alpha)\frac{\varrho(h)V^2}{2}S;$$

$$Z = c_z(M, \alpha)\frac{\varrho(h)V^2}{2}S,$$

where $\varrho(h)$ is the density of the atmosphere; S is the effective wing area; M is the Mach number, equal to the ratio of the flight velocity to the velocity of sound $a(h)$ at the given altitude; c_x, c_y and c_z are the aerodynamic coefficients.

Equations (I. 2) relate two essentially different groups of variables. The variables h, V, θ, ψ_c, x and m enter (I.2) together with their first derivatives and thus characterize the state of the system at any time t; the number of these variables is equal to the order of the system.

Such variables are generally known as p h a s e c o o r d i n a t e s . The variables α, β, γ_c, φ, and β_f enter (I.2) without their derivatives and thus act as free variables. They may be defined quite arbitrarily as certain functions of time, and they determine the solution of the system (the behavior of the controlled system) for a given initial state h_0, V_0, θ_0, ψ_{c0}, m_0. Variables of this kind are known as c o n t r o l e l e m e n t s or c o n t r o l s .

The classification of the variables into phase coordinates and control elements is closely linked with the particular choice of the mathematical model of the controlled system. In some problems the mathematical model (system (I. 2)) provides an insufficiently accurate description of the actual behavior of the aircraft and it can be improved by supplementing it with the equation of the angular motion of the aircraft about its center of mass. In this system, the variables α, β and γ_c become phase coordinates, and the rudder deflection angles assume the role of control elements. On the other hand, in some problems, certain phase coordinates may be upgraded to the status of control elements without any detrimental effects; this would involve dropping the corresponding differential equations from the mathematical model. This approach was actually applied by some authors /4, 8/ in solving the problems of powered ascent of aircraft, when the trajectory inclination angle θ was used as a control element.

In practical problems, besides differential equations we have to consider further a variety of constraints and conditions on the variables which stem from the particular properties of the controlled system. The following typical constraints are imposed on aircraft flying in the denser atmospheric layers: the altitude $h \geqslant 0$, the angle of attack $\alpha_{min} \leqslant \alpha \leqslant \alpha_{max}$, the dynamic head $q = {}^1\!/_2 \varrho V^2 \leqslant q_{max}$, the total overload $N = \dfrac{1}{a}\ \sqrt{X^2(h,V,\alpha) + Y^2(h,V,\alpha)} \leqslant$ $\leqslant N_{max}$, the surface temperature $T_w(h,\ V,\ \alpha) \leqslant T_{w\,max}$. Certain additional conditions are also imposed on the initial and the final state of the object. Thus, a vehicle is intended to transport some payload from the point of origin h_0, x_0 on the ground, where it was at rest $(V_0 = 0)$, to a circular orbit at altitude $h_1 (\theta_1 = 0,\ V_1 = V_{cir})$. The different control programs satisfying all these requirements will be referred to as a d m i s s i b l e a l t e r n a t i v e s .

The set of all admissible control programs D may be identified with the set M in the general formulation. We will consider i n t e g r a l f u n c t i o n a l s defined over the set D: these are integrals taken between certain time limits over some functions of the phase coordinates and control elements or functions of initial and final states, or linear combinations of these states. Examples of such functionals for aircraft are the flight range $\int_{t_0}^{t_1} V \cos \theta\, dt$, the flight duration $t_1 - t_0$, fuel consumption $m_0 - m_1$, the final altitude h_1, the final velocity V_1, etc.

We are thus dealing with the following general problem.

Consider a system of differential equations describing a controlled system,

$$\dot{y} = f(t,\ y,\ u), \qquad\qquad\qquad (I.\ 3)$$

where $y = (y^1, y^2, \ldots, y^n)$ is the n-dimensional vector of the phase coordinates, $u = (u^1, u^2, \ldots, u^r)$ is the r-dimensional vector of the control elements.

The variables y, u for every t may take on certain values from some set $V(t)$ defined by the additional constraints

$$(y, u) \in V(t). \tag{I.4}$$

The projection of this set onto the space of the phase coordinates Y at every t will be designated $V_y(t)$, $y \in V_y(t)$, and its cross section for any t and y will be designated $V_u(t, y)$, $u \in V_u(t, y)$. Expression (I.4) incorporates, in particular, the boundary conditions

$$y_0 \in V_y(t_0), \quad y_1 \in V_y(t_1),$$

where $y(t_0) = y_0$, $y(t_1) = y_1$.

In the set D of the pairs of vector functions $y(t)$, $u(t)$ satisfying the above conditions, find a sequence $\{\bar{y}_s(t), \bar{u}_s(t)\}$ over which the functional

$$I = \int_{t_0}^{t_1} f^0(t, y, u)\,dt + F(y_0, y_1) \tag{I.5}$$

goes to its exact lower bound in the set, $I(\bar{y}_s(t), \bar{u}_s(t)) \to \inf_D I$. In particular, if we assume that the least value of the function I is attained on some element $\bar{y}(t)$, $\bar{u}(t)$ of class D, our problem is to find this element:

$$I(\bar{y}(t), \bar{u}(t)) = \inf_D I. \tag{I.6}$$

Let us now consider briefly the generally accepted methods of solving problems of this kind.

Pontryagin's maximum principle

Let, for simplicity, the set V_u be constant (independent of t and y), the boundary conditions t_0, t_1, y_0, y_1 fixed, the set $V_y(t)$ for any $t \in (t_0, t_1)$ coincide with the space Y (there are no constraints on the phase coordinates), and $F(y_0, y_1) = 0$.

To solve the problem, we define a function of $(2n + r + 2)$ variables (the problem Hamiltonian)

$$H(t, \psi_0, \psi, y, u) = \sum_{i=1}^{n} \psi_i f^i(t, y, u) + \psi_0 f^0(t, y, u), \tag{I.7}$$

where $\psi = (\psi_1, \psi_2, \ldots, \psi_n)$ is an n-dimensional vector, the first term on the right is a scalar product of n-dimensional vectors. The so-called adjoint system of n differential equations is then added to the starting equations:

$$\dot{\psi} = -\frac{\partial H}{\partial y}. \tag{I.8}$$

The maximum principle is expressed by the following theorem.

T h e o r e m I. 1.   Let $(\bar{y}(t),\ \bar{u}(t)) \in D$ be a solution of the problem, i. e., the point of D minimizing the functional I.   Then there exists a vector $(\psi_0,\ \psi(t))$ which is not identically zero, where $\psi(t)$ is the solution of (I. 8) for $y=\bar{y}(t),\ u=\bar{u}(t),\ \psi_0$ is a nonpositive constant, such that for all $t \in [t_0,\ t_1]$ the function $H(t,\ \bar{y}(t),\ \psi_0,\ \psi(t),\ u)$ attains its absolute maximum on V_u for $u=\bar{u}$:

$$H\left(t,\bar{y}(t),\psi_0,\psi(t),\bar{u}(t)\right) = \sup_{u \in V_u} H\left(t,\bar{y}(t),\psi_0,\psi(t),u\right) = \mu(t) \tag{I. 9}$$

and the function $\mu(t)$ is continuous over $[t_0, t_1]$.

The proof of this fundamental theorem will be found in /6, 2, 3/.

Equations (I. 8)—(I. 9) are the necessary conditions for an optimum solution $\bar{y}(t),\ \bar{u}(t)$.   Together with equations (I. 3), they constitute a system of ordinary differential equations of order $2n$, closed by the finite relation (I. 9); on a given segment $[t_0,\ t_1]$ the solution of this system $(y(t),\ \psi(t),\ u(t))$ should satisfy $2n$ boundary conditions which constrain the phase path $y(t)$ to pass through given points $y_0,\ y_1$ at times t_0 and t_1 , respectively.

In other words, the maximum principle reduces the optimization problem to a boundary-value problem for a system of ordinary differential equations.

The solution of this boundary-value problem, in general, may prove to be different from the sought optimum solution, since the equations of the maximum principle provide only the necessary conditions of optimality. However, if we can be certain that, first, the minimum of the functional exists (in the set D) and, second, the solution of the boundary-value problem is unique, the solution obtained by this method is indeed the optimum solution.

The maximum principle generalizes the classical optimum conditions — the Euler—Lagrange equations and Weierstrass's condition — to the case of a closed and bounded set V_u (these are the conditions mostly encountered in practice).   If the set V_u coincides with the entire space U , we have from (I. 9)

$$H_u(t,\ \bar{y},\ \psi,\ \bar{u})=0, \tag{I. 10}$$

where $H_u=(H_{u^1},\ H_{u^2},\ \ldots,\ H_{u^r})$ is the vector of the partial derivatives of the function H with respect to the components of the vector u .   Equations (I. 3), (I. 8), and (I. 10) form a system of canonical Euler—Lagrange equations for the functions $y(t),\ \psi(t),\ u(t)$.   The functions $\psi_0,\ \psi_1(t),\ \psi_2(t),\ \ldots,\ \psi_n(t)$ coincide in this case with the Lagrange multipliers, and relation (I. 9) coincides with Weierstrass's necessary condition.   Indeed, from the definition of the supremum, (I. 9) implies that

$$H(t,\ \bar{y},\ \psi,\ \bar{u})-H(t,\ \bar{y},\ \psi,\ u) \geqslant 0, \tag{I. 11}$$

for any $u \in V_u$.

In general, when the set V_u is closed, the point at which the function $H(u)$ reaches its maximum need not be a stationary point of the function, so that the Euler—Lagrange equations (I. 10) are not always the necessary conditions of optimum.

Note that various transformations mapping the set V_u onto some open set, which are widely used by various authors (e. g., /7—9/) in conjunction with the apparatus of variational calculus, lead to the equations of the maximum principle (I. 3), (I. 8), (I. 10), if their application is valid.

Example I.1. As an example, let us consider the problem of the shortest path between two given points A and B. The set D of all the admissible paths is limited by additional conditions requiring single-valuedness of $y(x)$, $z(x)$ on (x_A, x_B). The functional and the equations describing the set D then take the form

$$I = \int_{x_A}^{x_B} \sqrt{1 + u^2 + v^2}\, dx; \qquad (\text{I.12})$$

$$\dot{y} = \frac{dy}{dx} = u; \quad \dot{z} = \frac{dz}{dx} = v; \qquad (\text{I.13})$$

$$y(x_A) = y_A; \quad y(x_B) = y_B; \qquad (\text{I.14})$$

$$z(x_A) = z_A; \quad z(x_B) = z_B. \qquad (\text{I.15})$$

Here the rectangular coordinates y and z are the phase coordinates of the problem, and the derivatives of the functions $y(x)$ and $z(x)$ are the control elements.

The vector ψ should also have two components. We designate them ψ_y and ψ_z, respectively: $\psi = (\psi_y, \psi_z)$.

By /6/, we have

$$H(x, \psi_0, \psi_y, \psi_z, y, z, u, v) = \psi_y u + \psi_z v + \psi_0 \sqrt{1 + u^2 + v^2},$$

i.e., the function H depends only on the vector ψ and the control elements u, v. Equations (I.8) thus take the form

$$\dot{\psi}_y = 0; \quad \dot{\psi}_z = 0,$$

whence it follows that the components of the vector ψ are constant:

$$\psi_y = \text{const}; \quad \psi_z = \text{const}.$$

Let us investigate the maximum of H with respect to u. For $\psi_0 = 0$, H has a maximum only for $\psi_y = \psi_z = 0$. The trivial solution $\psi_0 = \psi_y = \psi_z = 0$ does not satisfy the maximum principle, and we therefore have to take $\psi_0 \neq 0$. For any given x, $\psi_0 < 0$, ψ_y, ψ_z, the function $H(u, v)$ has a single point of maximum, $\bar{u}, \bar{v}$, which is independent of x.

Using equations (I.8), we find that the shortest path is the line of constant slope, i.e., a straight line; since only one straight line can be passed through two given points, the shortest path is unique. Since the shortest path between two points certainly exists, we conclude that our solution indeed gives the shortest path — the straight line joining the points A and B.

Bellman's dynamic programming method

Consider a particular case of the general problem, when the initial values of t and y are fixed, the final value of t is given, but no constraints are imposed on the final value of y. Moreover, there are no constraints on the phase vector in the entire interval (t_0, t_1), so that the set $V_y(t)$ is an open domain.

A suitable example is the optimal homing of a liquid-propellant rocket on a free target in vacuum. Constant engine thrust is assumed, but its direction can be altered at will. If the differences in gravitational acceleration on the two objects are ignored, the homing equations relative to the free target may be written in the form

$$\left.\begin{aligned}\dot{\vec{r}} &= \vec{V}; \\ \dot{\vec{V}} &= \vec{p}\,a(t),\end{aligned}\right\} \qquad (I.16)$$

where $\vec{r}, \vec{V}$ are the radius-vector and the velocity vector, $\vec{p}$ is the unit vector in the direction of the thrust (a control element), $a(t)$ is the thrust-produced acceleration (a known function of time). The initial position $\vec{r}_0$ and the initial velocity $\vec{V}_0$ of the homing rocket are known. The problem is to minimize the square of the distance to the target r_1^2 at the final time $t=t_1$.

In tackling problems of this kind, Bellman's method uses a self-evident optimum principle /1/.

Optimal behavior has the property that, whatever the initial state and the initial solution, subsequent solutions constitute an optimum in relation to the state ensured by the original solution. In other words: any component of an optimal process is an optimal process in relation to the current (instantaneous) state.

The sought optimal solution and the minimum value of the functional clearly depend only on the initial time t and the initial state y. We thus have a scalar function $S(t, y)$ defining the minimum value of the functional for given t, y, which are treated as the initial values of the problem. In our example, this scalar function is the square of the minimum distance to the target at the end of the homing run, assuming that the homing missile was launched at the time t from point $\vec{r}$ with velocity $\vec{V}$. The function $S(t, \vec{r}, \vec{V})$ in our example is thus defined as

$$S(t_1, \vec{r}, \vec{V}) \equiv r^2. \qquad (I.17)$$

The optimum principle is expressed by the following functional equation:

$$S(t, y_0) = \inf_{\{u\}} \left[\int_t^\tau f^0(t, y, u)\,dt + S(\tau, y_\tau) \right], \qquad (I.18)$$

where $\{u\}$ is the set of the admissible values of the control elements; τ is some time from the interval $[t, t_1]$; $y(t)$ is the solution of (I.3) in the time interval $[t, \tau]$ with the control element $u(t)$ and the initial conditions t, y; $y_\tau = y(\tau)$ is the value of this solution for $t=\tau$. Allowing τ to go to t and assuming continuity and differentiability of $S(t, y)$, we readily find the following partial differential equation:

$$\frac{\partial S}{\partial t} + \inf_{u \in V_u} \left\{ \frac{\partial S}{\partial y} f(t, y, u) + f^0(t, y, u) \right\} = 0, \qquad (I.19)$$

with boundary condition (I. 17).   When this equation is solved, we obtain
the function $S(t, y)$; the optimum control element at any point (t, y) is found
by minimizing the expression in braces.

It is readily seen that in this way we not only find the optimum program
for given initial values t_0, y_0, but in fact solve a more general problem of
the optimum behavior for any pair of values t, y, treated as the initial
values.   This type of solution is known as optimum control
synthesis.

In our particular example, Bellman's equation (I. 19) takes the form

$$\frac{\partial S}{\partial t} + \frac{\partial S}{\partial \vec{r}} \, \vec{V} + \inf_{\vec{p}} \left(\frac{\partial S}{\partial \vec{V}} \, \vec{p} a \, (t) \right) = 0. \tag{I. 20}$$

Seeing that $a(t) \geqslant 0$ and the scalar product $\dfrac{\partial S}{\partial \vec{V}} \, \vec{p}$ is minimum when the
unit vector $\vec{p}$ is antiparallel to $\dfrac{\partial S}{\partial V}$,

$$\vec{p} = - \frac{\partial S}{\partial \vec{V}} \left/ \left| \frac{\partial S}{d\vec{V}} \right| \right. , \tag{I. 21}$$

we rewrite (I. 20) in the form

$$\frac{\partial S}{\partial t} + \frac{\partial S}{\partial \vec{r}} \, \vec{V} - \left| \frac{\partial S}{\partial \vec{V}} \right| a \, (t) = 0. \tag{I. 22}$$

This is a nonlinear partial differential equation of first order.   Its solution
with boundary condition (I. 17) gives a certain function $\tilde{S}(t, \vec{r}, \vec{V})$.   Expression
(I. 21) then defines the field of optimum control elements, i. e., the optimum
direction of the thrust vector for any state $t, \vec{r}, \vec{V}$.

Like the maximum principle, Bellman's equation has its classical
analog — the Hamilton—Jacobi equation of classical mechanics which differs
from (I. 19) in that u is obtained from the condition

$$\frac{\partial}{\partial u} \left\{ \frac{\partial S}{\partial y} \, f(t, y, u) + f^0(t, y, u) \right\} = 0, \tag{I. 23}$$

and not by minimizing the expression in braces.

Equation (I. 19) is evidently more general and it essentially coincides
with the Hamilton—Jacobi equation in cases when the given expression attains
its minimum at a unique stationary point.

Both methods — the maximum principle and Bellman's dynamic
programming method — are widely used because they are organically linked
with the particular features of modern applied problems and allow in a
straightforward manner for the physical constraints imposed on the various
elements.   A wide range of optimum control problems have been solved by
these methods.

There exists another extensive class of problems (including typical
flight dynamics problems), however, which are very difficult and often
impossible to tackle by these methods.

Let us consider some simple examples of this kind.

Example I. 2.   Consider the problem of the optimal conditions of the vertical ascent of a rocket launched from the surface of the Earth, which is required to reach a given altitude with minimum fuel consumption (or, equivalently, to reach a maximum altitude with a given fuel charge) /5/.
The equations of motion of the rocket are

$$\dot{h}=V; \tag{I.24}$$

$$\dot{V}=-\frac{X(h,V)}{m}-g(h)+P/m; \tag{I.25}$$

$$\dot{m}=-\frac{1}{c}P. \tag{I.26}$$

Here, h, V, m are respectively the altitude, the velocity, and the mass of the rocket; $X(h, V)$ is the drag; $g(h)$ is the gravitational acceleration; P is the engine thrust, controllable between the limits $0 \leqslant P \leqslant P_{max}$; c is the nozzle velocity of the combustion gases (constant). Let h_0, V_0, m_0, t_0 be the given initial values of the altitude, the velocity, the mass, and time; h_1, t_1 are the final altitude and time.

A characteristic feature of this problem is the linear dependence of the right-hand sides of equations (I. 24)−(I. 26) on the controlled thrust P.

Let us analyze the problem using Pontryagin's maximum principle.
Let

$$\left.\begin{array}{c} H(\psi_1,\psi_2,\psi_3,h,V,m,P)=\psi_1 V +\psi_2\left(-\frac{1}{m}X(h,V)-g(h)+\frac{P}{m}\right)-\psi_3\frac{1}{c}P, \\[2ex] \dot{\psi}_1=-\frac{\partial H}{\partial h}; \quad \dot{\psi}_2=-\frac{\partial H}{\partial V}; \quad \dot{\psi}_3=-\frac{\partial H}{\partial m} \end{array}\right\} \tag{I.27}$$

be the Hamiltonian and the adjoint system of equations, respectively. The optimum solution is then selected from among the solutions of the system of differential equations (I. 24)−(I. 27) closed by the additional relation

$$H(\psi_1,\psi_2,\psi_3,h,V,m,\overline{P})=\max_{0\leqslant P\leqslant P_{max}} H(\psi_1,\psi_2,\psi_3,h,V,m,P). \tag{I.28}$$

The sought solutions should satisfy the following boundary conditions:
for $t=t_0$,

$$h=h_0; \ V=V_0; \ m=m_0;$$

for $t=t_1$,

$$h=h_1; \ \psi_2(t_1)=0; \ H(\psi_1,\psi_2,\psi_3,h,V,m,\overline{P})_{t=t_1}=0$$
$$\psi_3(t_1)>0.$$

Let us consider (I. 28) in more detail. Seeing that the function H depends linearly on the controlled thrust, we readily conclude that depending on the sign of the coefficient before P (the switching function),

$$M = \psi_2/m - \psi_3/c, \qquad\qquad (I.29)$$

this equation is satisfied by

$$
\begin{array}{ll}
1)\ \overline{P}=P_{\max} & \text{for } M>0 \\[4pt]
2)\ \overline{P}=0 & \text{for } M<0 \\[4pt]
3)\ M=0 &
\end{array}
\right\} \qquad\qquad (I.30)
$$

In case 3, H is evidently independent of P, so that any $\overline{P}$ may be chosen from $[0,\ P_{\max}]$.

In our example, case 3 in (I.30) may be identically satisfied over some time interval The corresponding solution in the theory of optimal processes is known as the singular control solution.

The corresponding control function is obtained by making the finite equation $M=0$ consistent with system $(I.24)-(I.27)$.

Setting the total time derivative of M equal to zero,

$$\dot{M} = \frac{\dot{\psi}_2}{m} - \frac{\psi_2 \dot{m}}{m^2} - \frac{\dot{\psi}_3}{c} = 0, \qquad\qquad (I.31)$$

and inserting the right-hand sides of $(I.24)-(I.27)$ together with the equality $H=0$, we obtain a new finite relation, which is independent of P:

$$N = \frac{1}{m}\left[X\left(\frac{1}{V}-\frac{1}{c}\right) - X_V \right] + \frac{g}{V} = 0. \qquad\qquad (I.32)$$

Repeating the same operation for (I.32), we obtain a linear equation for P, from which the sought control function is found.

The construction of the optimum program generally involves the solution of a boundary-value problem for $(I.24)-(I.28)$. As we have seen above, the rationale of the procedure based on the necessary conditions only is twofold: first, the existence of the sought optimum program must be established and, second, the uniqueness of the solution satisfying these necessary conditions should be proved. The existence of an optimum program for the problem being considered (the right-hand sides of the equations $(I.24)-(I.26)$ are linear functions of the control elements) follows from the general considerations of the theory of optimal processes. The uniqueness of the solution of a boundary-value problem for a nonlinear system of differential equations, on the other hand, is by no means certain. The existence of singular control markedly aggravates the situation. Indeed the singular control program is the set of switching points $M=0$, so that at each of these points we may take both $P=0$ and $P=P_{\max}$ (Figure I.1).

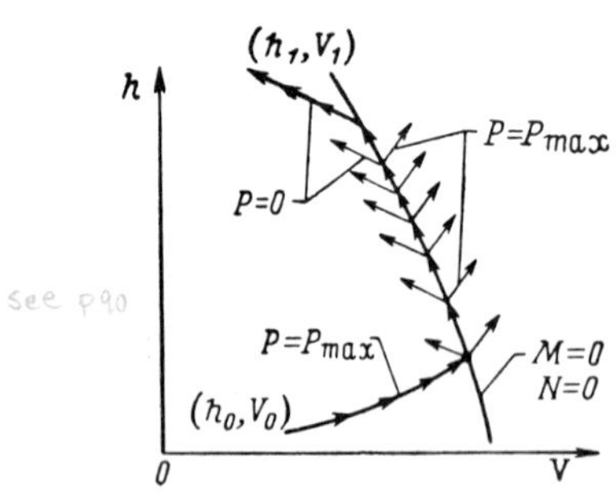

FIGURE I.1

see p90

Not every trajectory obtained in this way need be consistent with the boundary conditions, and there are in fact examples showing that there may be infinitely many such trajectories.

We thus see that complete solution of problems of this kind requires sufficient or, at least, additional necessary conditions of optimum, which would enable us to eliminate all the incompatible solutions.

The dynamic programming method in this case leads to a nonlinear partial differential equation

$$\frac{\partial S}{\partial t} + \frac{\partial S}{\partial h} V + \frac{\partial S}{\partial V}\left(-\frac{1}{m} X(h,V) - g(h)\right) +$$

$$+ \max_{0 \leqslant P \leqslant P_{max}} \left[\left(\frac{\partial S}{\partial V}\frac{1}{m} - \frac{\partial S}{\partial m}\frac{1}{c}\right) P\right] = 0,$$

or

$$\frac{\partial S}{\partial t} + \frac{\partial S}{\partial h} V + \frac{\partial S}{\partial V}\left(-\frac{1}{m} X(h,V) - g(h)\right) + \varkappa(M) = 0, \tag{I. 33}$$

where

$$\varkappa(M) = \begin{cases} MP & \text{for} \quad M = \dfrac{\partial S}{\partial V}\dfrac{1}{m} - \dfrac{\partial S}{\partial m}\dfrac{1}{c} > 0, \\[2mm] 0 & \text{for } M < 0. \end{cases}$$

No sufficiently general theorems of the existence of solutions for these equations (with a non-differentiable left-hand part) are available; nor are there regular methods for their solution (numerical or otherwise).

We are thus faced with a contradiction: in those cases when the additional necessary and sufficient conditions of optimum are required for constructive purposes, they are in fact inapplicable. New effective optimum conditions of singular control thus have to be found.

Example I. 3. Let us consider the same problem as in Example I. 1 with one difference: the engine is not throttled, and it is either on or off (the thrust P may only take on two values, 0 and P_{max}). In this case, as is readily seen, the equation $M = 0$ is no longer compatible with equations (I. 24) − (I. 27) of the maximum principle, so that the singular control solution is not included among the allowed solutions of these equations. However, even if a unique solution has been obtained for the boundary-value problem of equations (I. 24) − (I. 27), we can by no means be certain that this solution is the optimum, as we do not know that the optimum solution exists among the admissible solutions. The problem is solved in this case by a minimizing sequence, providing a so-called sliding control program. It will be shown in what follows that a typical solution of this problem is a succession of thrust programs with infinitely increasing thrust switching frequency around the line $M = 0$, $N = 0$.

We are thus faced with a new problem of finding appropriate minimizing sequences.

Problems of this class can be replaced by another problem, for which
the optimum solution is known to exist. However, the minimizing sequence
of the original problem corresponds to the singular control solution of the
new problem, and in the final analysis the sliding control problem reduces
to a singular control problem. Bellman's equation for this problem
precisely coincides with equation (I. 33), with all its consequences.

E x a m p l e I. 4. Suppose that the thrust in Example I. 2 does not have
a maximum value (this is a typical situation in the mechanics of space
flight). In this case, as is readily seen, both the function H and the left-
hand side of Bellman's equation are infinite for positive M, so that neither
the maximum principle nor the dynamic programming method are
applicable in their original form. Moreover, with arbitrary boundary
conditions, the admissible solutions of the problem do not include the
optimum solution, so that we have to construct a minimizing sequence,
although of a different type: a sequence allowing indefinite growth of the
thrust. This is a typical feature of systems with unbounded linear control
elements. In addition to these fundamental difficulties, there are
difficulties of secondary importance associated with the actual numerical
solution of the problems. This remark primarily applies to Bellman's
method. How are we to solve numerically equation (I. 19), say? Consider
the solution by the grid method, when we have to compute the partial
derivatives of the function $S(t_1, y)$ at certain tabular points (the function is
known for $t=t_1$).

From equation (I. 19) we then find the values of $\partial S/\partial t$ at the same tabular
points and approximate to the values of S for $t = t_1 - \Delta$ using the equality

$$S(t_1 - \Delta) = S(t_1, y) - \frac{\partial S(t, y)}{\partial t} \Delta.$$

The same procedure is then repeated for $S(t_1 - \Delta,\ y)$, etc., until we
reach t_0. Consider the case of a midcourse maneuver in space. Then
$\vec{r}, \vec{V}$ are three-dimensional vectors, and $S(t, \vec{r}, \vec{V})$ is a function of seven
variables. The numerical grid consists of ten values of each variable for
every t. It thus comprises 10^6 points, at which the function S should be
computed and the results stored. Such a tremendous volume of data
cannot be stored in the immediate-access memory of most computers.

And yet, this is a very modest number considering the required
accuracy. If the grid spacing is reduced only by a factor of 2, this number
will increase by a factor of $2^6 = 64$, so that even the most optimistic
forecasts of the future development of computers can hardly catch up with
this "growth". Suppose that equation (I. 19) has been solved; it is still not
clear, however, how we are to store and use the solution, which is a
function of seven variables defined by its numerical values. Certain
techniques can be applied in order to reduce this "curse of dimensionality",
to borrow Bellman's expression, but it cannot be eliminated completely, so
that at the present stage nonlinear problems of third or fourth order
apparently constitute the limit as far as Bellman's method is concerned.

The maximum principle does not lead to such catastrophic effects,
and it is much more powerful and more promising in this respect. In view
of the difficulties involved in the solution of boundary-value problems, we
have to concentrate on developing alternative methods of optimal solution,

which will not require straightforward comparison of all the different alternatives satisfying the necessary optimality conditions. This is a highly important and as yet unsolved problem.

A new approach to the solution of the general problem, considered in the next section, opens new possibilities for overcoming these fundamental difficulties.

Optimum principle based on a reduction of the given problem to a trivial problem

Consider the problem of minimizing the functional

$$I(y(t),\ u(t)) = -\int_{t_0}^{t_1} R(t,\ y,\ u)\,dt + \Phi(y_0,\ y_1) \tag{I.34}$$

on the set E of pairs of vector functions $y(t)$, $u(t)$, which differs from the set D only in that $y(t)$ and $u(t)$ are not related by differential equations. This problem will be called a t r i v i a l   p r o b l e m. The following almost obvious optimum conditions apply to this problem. The functional I is minimized on the set E by the vector functions $\bar{y}(t)$, $\bar{u}(t)$ whose values at every interior point of the segment $[t_0,\ t_1]$ maximize the integrand $R(t,\ y,\ u)$ over the set $V(t)$ and the values at the integration limits t_0, t_1 minimize the function $\Phi(y_0,\ y_1)$ over the set $V_y(t_0) + V_y(t_1)$:

$$R(t,\ \bar{y},\ \bar{u}) = \sup_{(y,\ u)\in V(t),} R(t,\ y,\ u)\ t\in(t_0,\ t_1); \tag{I.35}$$

$$\Phi(\bar{y}_0,\ \bar{y}_1) = \inf_{\substack{y_0\in V_y(t_0)\\ y_1\in V_y(t_1)}} \Phi(y_0,\ y_1). \tag{I.36}$$

In other words, the problem of minimizing the functional over the set of functions $y(t)$, $u(t)$, i. e., over an infinite set of numbers, is thus reduced to finding the maximum of a function $R(t,\ y,\ u)$ of $n+r$ variables y^i, u^k for every given $t\in(t_0,\ t_1)$ and the minimum of the function $\Phi(y_0,\ y_1)$ of $2n$ variables.

These conditions, with slight reservations, are generalized to the case of a minimizing sequence $\{\bar{y}_s(t),\ \bar{u}_s(t)\}$.

This sequence satisfies the conditions

$$R(t,\ \bar{y}_s,\ \bar{u}_s) \to \sup_{(y,\ u)\in V(t)} R(t,\ y,\ u),\ t\in(t_0,\ t_1); \tag{I.37}$$

$$\Phi(\bar{y}_s(t_0),\ \bar{y}_s(t_1)) \to \inf_{\substack{y_0\in V_y(t_0)\\ y_1\in V_y(t_1)}} \Phi(y_0,\ y_1). \tag{I.38}$$

In view of the simplicity of these conditions, it is advisable to reduce the general variational problem with differential constraints to a trivial problem, i. e., we have to find the functions $R(t, y, u)$ and $\Phi(y_0, y_1)$ such that the solution of the trivial problem is also a solution of the original problem. This is the essence of the variational method considered in this book.

Let us present, without proof, the basic result on which the variational principle devolves. The functions R and Φ are sought in the following form:

$$R(t, y, u) = \sum_{i=1}^{n} \varphi_{y^i} f^i(t, y, u) - f^0(t, y, u) + \varphi_t; \qquad (\text{I. 39})$$

$$\Phi(y_0, y_1) = F(y_0, y_1) + \varphi(t_1, y_1) - \varphi(t_0, y_0), \qquad (\text{I. 40})$$

where φ_{y^i}, φ_t are the partial derivatives of some function $\varphi(t, y)$ which is continuous and differentiable over $V_y(t)$.

Let

$$\mu(t) = \sup_{y, u \in V(t)} R(t, y, u); \qquad (\text{I. 41})$$

$$m = \inf_{y_0 \in V_y(t_0),\ y_1 \in V_y(t_1)} \Phi(y_0, y_1). \qquad (\text{I. 42})$$

T h e o r e m I. 2. Consider a pair of vector functions $\bar{y}(t)$, $\bar{u}(t)$ from the set D and a function $\varphi(t, y)$, such that

1) $R\left[t, \bar{y}(t), \bar{u}(t)\right] = \mu(t);$ \hfill (I. 43)

2) $\Phi\left[\bar{y}(t_0), \bar{y}(t_1)\right] = m.$ \hfill (I. 44)

This pair $\bar{y}(t)$, $\bar{u}(t)$ minimizes the functional I . The functions $R(t, y, u)$ and $\Phi(y_0, y_1)$ mentioned above thus have the form (I. 39), (I. 40) and are expressed in terms of the function $\varphi(t, y)$ of the $n+1$ arguments t, y^i, $i = 1, 2, \ldots, n$, which satisfies the conditions of the theorem.

In a more general case, when our problem is to find a minimizing sequence, Theorem I. 2 is formulated in a somewhat different form.

T h e o r e m I. 3. Consider a sequence $\{\bar{y}_s(t), \bar{u}_s(t)\}$ from D and a function φ, such that

1) $R\left[t, \bar{y}_s(t), \bar{u}_s(t)\right] \to \mu(t), t \in (t_0, t_1);$ \hfill (I. 45)

2) $\Phi\left[\bar{y}_s(t_0), \bar{y}_s(t_1)\right] \to m;$ \hfill (I. 46)

3) $\mu(t)$ is piecewise-continuous and m is a finite number;
4) the sequence $R[t, \bar{y}_s(t), \bar{u}_s(t)]$ is bounded.

Then this sequence minimizes the functional I on the set D . Conditions 1 and 3 may be replaced by a more general, though less obvious, condition

$$\int_{t_0}^{t_1} R[t, \bar{y}_s(t), \bar{u}_s(t)]\, dt \to \int_{t_0}^{t_1} \mu(t)\, dt. \qquad (\text{I. 47})$$

The main difficulty associated with the application of this principle is the definition of the function $\varphi(t, y)$.

The definition of φ essentially determines the method of solving the problem. We will see at a later stage that one of the possible definitions of φ leads, in particular, to the equations of Pontryagin's maximum principle as the necessary conditions for the maximum of the function R, while another method leads to Bellman's equations. These two alternatives, however, do not exhaust all the different possibilities.

The arbitrariness inherent in the definition of φ may be put to work so as to devise the best procedure for each particular problem.

Let us consider two simple examples which illustrate this point.

E x a m p l e I. 5. Minimize the functional

$$I = \int\limits_0^1 (y^2 - u)\, dt$$

subject to the constraints

$$\dot{y} = u; \quad |u| \leqslant 1; \quad y(0) = y(1) = 0.$$

Here y and u are scalar functions.

The function $R(t, y, u)$ has the form

$$R = \varphi_y u - y^2 + u + \varphi_t. \tag{I.48}$$

(The function Φ is a priori minimized by Theorems I. 2 — I. 3, since y_0, y_1 are fixed numbers.) The function φ is defined so as to make R independent of u. To this end, it suffices to take

$$\varphi_y = -1, \tag{I.49}$$

whence

$$\varphi = -y + C(t),$$

where $C(t)$ is any differentiable function.

Thus,

$$R = -y^2 + C_t, \tag{I.50}$$

and this function has a maximum for $y = 0$ (with any u).

Setting $\bar{y}(t) = 0$ in the equation $\dot{y} = u$, we find $\bar{u}(t) \equiv 0$. Since $\bar{y}(t)$ satisfies the boundary conditions, the pair $\bar{y}(t) = 0$, $\bar{u}(t) = 0$ is the sought solution minimizing the functional. Note that we have obtained the minimizing solution without solving any boundary-value problems.

For comparison, let us consider another definition of φ, which leads to Bellman's equation. According to this approach, we find the maximum of the function R with respect to u for any t and y and set it identically equal to zero:

$$\sup_{|u| \leqslant 1} R(t, y, u) = |\varphi_y + 1| - y^2 + \varphi_t = 0.$$

This is a fairly complicated nonlinear partial differential equation (with a nondifferentiable left-hand side). The first technique is thus much simpler.

Example I.5. Under the same constraints, minimize the functional

$$I = \int_0^1 (y^2 - u^2)\, dt.$$

The function R has the form

$$R = \varphi_y u - y^2 + u^2 + \varphi_t. \qquad (I.51)$$

We take $\varphi_y = 0$; $\varphi_t = 0$. The function R, for any t and y, thus has two maxima, $u = \pm 1$. Its maximum with respect to y, as before, corresponds to $\bar{y} = 0$. We thus have

$$\bar{y}(t) = 0; \quad \bar{u}_1(t) = +1; \quad \bar{u}_2(t) = -1.$$

It is readily seen that neither of the two solutions maximizing the function R satisfies the differential equation $\dot{y} = u$. However, the existence of three points at which R has maxima enables us to construct a sequence of solutions of the equation $\dot{y} = u$ such that R goes to a maximum for every t,

$$R(t,\ \bar{y}_S(t),\ \bar{u}_S(t)) \rightarrow \mu(t) = 1.$$

This sequence is constructed as shown in Figure I.2.

As the thrust switching frequency indefintely increases $(S \rightarrow \infty)$, we have $y_S(t) \rightarrow 0$. For any S we have $\bar{u}_S^2(t) = 1$.

This sequence provides conditional sliding solution with zero closeness function $y(t) \equiv 0$ and basis controls $u_{1,2} = \pm 1$.

The above examples have their own characteristic features, as the previously considered problems. In each case, the definition of φ took into consideration the particular features of the problem, and thus led to an effective solution of the problem by a method entirely different from the traditional techniques.

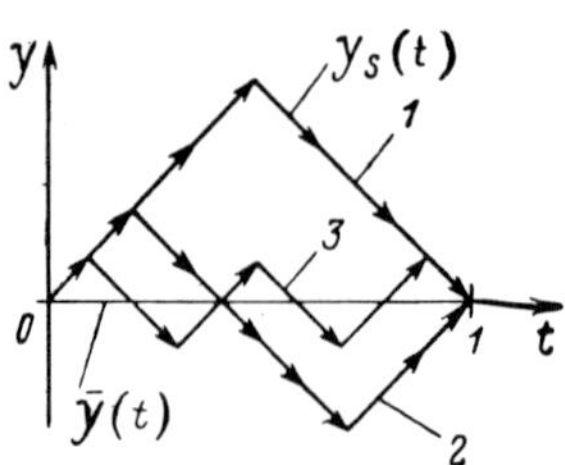

FIGURE I.2

Bibliography for Introduction

1. Bellman, R. Dynamic Programming. — Princeton Univ. Press. 1957.
2. Boltyanskii, V.G. Matematicheskie metody optimal'nogo upravleniya (Mathematical Methods of Optimal Control). — Nauka. 1966.
3. Dubovitskii, A.Ya. and A.A. Milyutin. Zadachi na ekstremum pri nalichii ogranichenii (Extremum Problems with Constraint). — Zhurnal Vychislitel'noi Matematiki i Matematicheskoi Fiziki, 5, No. 3. 1965.

4. O s t o s l a v s k i i, I. V. and I. V. S t r a z h e v a. Dinamika poleta.
   Traektorii letatel'nykh apparatov (Flight Dynamics. Aircraft
   Trajectories). — Oborongiz. 1965.
5. O k h o t s i m s k i i, D. E. K teorii dvizheniya roket (The Theory of
   Rocket Motion). — Prikladnaya Matematika i Mekhanika, **10**, No. 2.
   1946.
6. P o n t r y a g i n, L. S. et al. Matematicheskaya teoriya optimal'nykh
   protsessov (Mathematical Theory of Optimal Processes). —
   Fizmatgiz. 1961.
7. T r o i t s k i i, V. A. O variatsionnykh zadachakh optimizatsii protsessov
   upravleniya (Variation Optimization Problems for Control
   Processes). — Prikladnaya Matematika i Mekhanika, **26**, No. 1.
   1962.
8. M i e l e, A. General Variational Theory of the Flight Paths of Rocket
   Powered Aircraft, Missiles and Satellite Carriers. — Astronautica
   Acta. 1958.
9. V a l e n t i n e, F. A. The Problem of Lagrange with Differential
   Inequalities as Added Side Conditions. — A dissertation. The
   University of Chicago. 1937.

Chapter I

ELEMENTS OF THE THEORY

§ 1.1. STATEMENT OF THE PROBLEM

Our problem is to minimize the functional

$$I = \int_{t_0}^{t_1} f^0(t, y, u)\, dt + F(y_0, y_1) \tag{1.1}$$

on a set D of the pairs of vector functions $y(t)$, $u(t)$ satisfying the following conditions:

a) The vector function $y(t) = [y^1(t), \ldots, y^n(t)]$ is defined over the segment $[t_0, t_1]$; its components $y^i(t)$ $(i=1, 2, \ldots, n)$ are continuous over $[t_0, t_1]$ and have a piecewise-continuous derivative; for any fixed $t \in [t_0, t_1]$, the vector $y(t)$ belongs to a given region $V_y(t)$ in an n-dimensional vector space.

b) The vector function $u(t) = [u^1(t), \ldots, u^r(t)]$ is defined over $[t_0, t_1]$; its components $u^j(t)$ $(j=1, 2, \ldots, r)$ are continuous everywhere in (t_0, t_1), except for a finite number of points, where they may have discontinuities of the first kind; for any $t \in (t_0, t_1)$, $y \in V_y(t)$, the vector $u(t)$ belongs to a given set $V_u(t, y)$.

The function $F(y_0, y_1)$ is continuous for all y_0, y_1, where

$$y_0 = y(t_0), \quad y_1 = y(t_1). \tag{1.2}$$

The conditions imposed on $y(t)$ and $u(t)$ define a set $V(t)$ of admissible values of combinations of $n+r$ numbers (y^i, u^j) for every $t \in (t_0, t_1)$, a region V_y of the admissible values of (t, y) in the $(n+1)$-dimensional (t, y) space, and a set V of admissible combinations of $n+r+1$ numbers (y^i, u^j, t) $(i=1, 2, \ldots, n, j=1, 2, \ldots, r)$. In addition to the above conditions, the pair of vector functions $y(t)$, $u(t)$ should also satisfy the system of n differential equations

$$\dot{y} = f(t, y, u), \tag{1.3}$$

where $f = (f^1, f^2, \ldots, f^n)$.

The functions $f^i(t, y, u)$, $i=0, 1, \ldots, n$, are defined and continuous for all t, y, u.

The vector $y = (y^1, y^2, \ldots, y^n)$ is generally known as the phase vector or the state vector, and its components are the phase coordinates. The vector $u = (u^1, u^2, \ldots, u^r)$ is known as the control vector, and its components are the control functions or controls. Formally the phase coordinates are distinguished from the control functions in that the differential equations

(1.3) contain derivatives of the phase coordinates and do not contain derivatives of the control functions. The sets $V_y(t_0)$ and $V_y(t_1)$ of the admissible values of y_0, y_1 constitute the boundary conditions.

The argument t may be identified with time, any ascending function, a phase coordinate, or some other parameter.

The problem is formulated as follows: in the set D of the pairs of functions $y(t)$, $u(t)$ find a pair $\bar{y}(t)$, $\bar{u}(t)$, which minimizes (maximizes) the functional I.

If the set D does not contain such a pair of functions, the problem is formulated in a slightly different form: find a sequence $\{\bar{y}_s(t), \bar{u}_s(t)\} \subset D$ such that for $S \to \infty$ the functional (1.1) goes over this sequence to its lower-bound value on the set D.

The sought pair of functions $\bar{y}(t)$, $\bar{u}(t)$ will be referred to as the optimum solution or the absolute minimizing solution, and the sought sequence of $\bar{y}_s(t)$, $\bar{u}_s(t)$ will be called a minimizing (optimizing) sequence.

We will mainly be concerned with minimizing a given functional. The problem of maximizing the functional can always be reduced to the minimization problem by a simple reversal of the functional sign.

§ 1.2. THE SUFFICIENT CONDITIONS OF AN ABSOLUTE MINIMUM

In this section we will prove a theorem which constitutes the basis for all the methods of solving the optimization problems described in this book. First, however, we have to prove one lemma.

Let the functional $I(v)$ be defined over some set M, $v \in M$. Let

$$\inf_{v \in M} I(v) = m. \tag{1.4}$$

Find the absolute minimum of I on M, i.e., find a sequence $\{\bar{v}_s\} \subset M$, such that

$$\lim_{S \to \infty} I(\bar{v}_s) = m. \tag{1.5}$$

We call this sequence a minimizing sequence and it is said to minimize the functional I over the set M. If there exists an element $\bar{v} \in M$ such that

$$I(\bar{v}) = m, \tag{1.6}$$

we may take $\bar{v}_s = \bar{v}$ ($S = 1, 2 \ldots$). In this case the problem reduces to finding the element $\bar{v} \in M$ for which the functional I attains its absolute minimum over the set M.

Consider a set $N \supset M$ over which a functional L is defined, so that $L(v) = I(v)$ for $v \in M$.

Lemma. Consider a sequence $\{\bar{v}_s\} \subset M$ satisfying the condition

$$\lim_{S \to \infty} L(\bar{v}_s) = l, \tag{1.7}$$

where

$$l = \inf L(v) \quad v \in N.$$

This sequence minimizes the functional I over M:

$$I(\bar{v}_s) \underset{S \to \infty}{\to} m.$$

P r o o f .   Let (1.7) hold true. By definition, $L(\bar{v}_S)=I(\bar{v}_S)$. We will now show that $l=m$. If this is not so, then from $M \subset N$ we have $m>l$ so that $m-l>\varepsilon>0$. Since $\{\bar{v}_S\} \subset M$, we have $I(\bar{v}_S)-l>\varepsilon$ for any S. This, however, contradicts (1.7). Hence $l=m$, and (1.5) follows from (1.7). Q. E. D.

This lemma enables us to replace the problem of minimizing a functional over a set M by an analogous problem over a larger set N. The "augmented" problem may prove to be simpler if the structure of the set N is relatively simple.

Consider a function $\varphi(t, y)$ which is continuous for all t, y and has continuous partial derivatives φ_t, $\varphi_y=(\varphi_y{}^1,\ldots, \varphi_y{}^n)$ for all t, y, except a finite number of sets $t=$ const in the (t, y) space. Now construct the functions

$$\Phi(y_0, y_1)=F(y_0, y_1)+\varphi(t_1, y_1)-\varphi(t_0, y_0); \tag{1.8}$$

$$R(t, y, u)=\varphi_y f(t, y, u)-f^0(t, y, u)+\varphi_t; \tag{1.9}$$

$$\mu(t)=\sup_{(y,u)\in V(t)} R(t, y, u). \tag{1.10}$$

Here φ_y is an n-dimensional vector function, and the first term on the right in (1.9) is a scalar product of n-dimensional vectors:

$$\varphi_y \cdot f=\sum_{i=1}^{n} \varphi_y{}^i f^i.$$

The vectors y and u on the right in (1.10) are assumed to be independent.

T h e o r e m   1.1. Consider a sequence $\{y_S(t), u_S(t)\} \subset D$. For this sequence to minimize the functional I over the set D, it is sufficient that there exists a function $\varphi(t, y)$ such that

1) $R[t, \bar{y}_S(t), \bar{u}_S(t)] \to \mu(t), \ t \in (t_0, t_1)^*;$ \hfill (1.11)

2) $\displaystyle \Phi(\bar{y}_{0S}, \bar{y}_{1S}) \underset{S \to \infty}{\to} \inf_{y_0 \in V_y(t_0), \ y_1 \in V_y(t_1);} \Phi(y_0, y_1)>-\infty.$ \hfill (1.12)

3) $\mu(t)$ is piecewise-continuous over $[t_0, t_1]$;

4) there exists a finite number Q such that for any $S<\infty$ and any $t \in (t_0, t_1)$

$$R(t, \bar{y}_S(t), \bar{u}_S(t))>Q.$$

R e m a r k . If the absolute minimizing solution $(\bar{y}(t), \bar{u}(t)) \in D$ exists, Theorem (1.1) takes the following form. Consider the pair $(\bar{y}(t), \bar{u}(t)) \in D$. A sufficient condition for the functional (1.1) to attain an absolute minimum on this pair is the existence of a function $\varphi(t, y)$, such that

1) $R(t, \bar{y}(t), \bar{u}(t))=\mu(t)$ \hfill (1.11*)

almost everywhere in (t_0, t_1);

2) $\displaystyle \Phi(\bar{y}_0, \bar{y}_1)= \inf_{y_0 \in V_y(t_0), \ y_1 \in V_y(t_1)} \Phi(y_0, y_1).$ \hfill (1.12*)

Conditions 3 and 4 in this case are satisfied automatically.

P r o o f . Here the set M of the previous lemma is identified with the set D of the pairs of vector functions $y(t), u(t)$. The set N is defined as the set E of the pairs of vector functions $y(t), u(t)$ which differs from D in the following two respects: first, the functions $y^i(t), (i=1, 2,\ldots,n)$ may have discontinuities of the first kind at a finite number of points of the segment $[t_0, t_1]$ and, second, the vector functions $y(t), u(t)$ are not related by differential equations (1.3). We define the following functional on E:

---

* Convergence in the measure is implied here: the measure of the set of points $t \in (t_0, t_1)$, where $\mu(t)-R_S>\varepsilon$, goes to zero for any given ε.

$$L(y(t), u(t)) = \Phi(y_0, y_1) - \int_{t_0}^{t_1} R(t, y(t), u(t)) dt. \qquad (1.13)$$

On D we have $L=l$. Indeed, using (1.3) and (1.9), we get

$$R = \dot{\varphi} - f^0, \quad (y(t), u(t)) \in D, \qquad (1.14)$$

where $\dot{\varphi}$ is the total derivative of the function φ in virtue of (1.3).

Inserting (1.14) in (1.13) and remembering that $y(t)$ are continuous functions for $u(t)) \in D$, we obtain $L=l$.

Suppose that there exists a function $\varphi(t, y)$ and a sequence $\{\bar{y}_S(t), \ \bar{u}_S(t)\} \in D$ satisfying conditions 1 through 4 of the theorem. We will show that in this case $\{\bar{y}_S(t), \bar{u}_S(t)\}$ is a minimizing sequence of the functional L over the set E, i. e.,

$$\lim_{S \to \infty} L\left(\bar{y}_S(t), \ \bar{u}_S(t)\right) - l = 0,$$

where

$$l = \inf_{E} L(y(t), u(t)). \qquad (1.15)$$

Since the functions $y(t)$ which belong to the set E may have discontinuities, the first and the second terms in expression (1.13) for L are independent. Therefore

$$l = \inf_{(y_0 y_1) \in V_y(t_0) + V_y(t_1)} \Phi(y_0, y_1) - \sup_{E} \int_{t_0}^{t_1} R(t, y(t), u(t)) dt. \qquad (1.16)$$

We have

$$\lim_{S \to \infty} L\left(\bar{y}_S(t), \ \bar{u}_S(t)\right) - l = \lim_{S \to \infty} \Phi\left(\bar{y}_{0S}, \ \bar{y}_{1S}\right) -$$

$$\left(- \inf_{(y_0, y_1) \in V_y(t_0) + V_y(t_1)} \Phi(y_0, y_1) + \sup_{E} \int_{t_0}^{t_1} R(t, y(t), u(t)) dt - \right)$$

$$- \lim_{S \to \infty} \int_{t_0}^{t_1} R\left(\bar{y}_S(t), \ \bar{u}_S(t)\right) dt. \qquad (1.17)$$

Using (1.11), (1.12), and Lebesgue's theorem on the limit of an integral, we write

$$\lim_{S \to \infty} L\left(\bar{y}_S(t), \ \bar{u}_S(t)\right) - l = \sup_{E} \int_{t_0}^{t} R(t, y(t), u(t)) dt -$$

$$- \lim_{S \to \infty} \int_{t_0}^{t} R\left(t, \bar{y}_S(t), \ \bar{u}_S(t)\right) dt =$$

$$= \sup_{E} \int_{t_0}^{t_1} R(t, y(t), u(t)) dt - \int_{t_0}^{t_1} (\lim_{S \to \infty} R\left(\bar{y}_S(t), \ \bar{u}_S(t)\right) dt =$$

$$= \sup_E \int_{t_0}^{t_1} R(t,\, y(t),\, u(t))\, dt - \int_{t_0}^{t_1} \mu(t)\, dt. \tag{1.18}$$

Suppose that our proposition is not true, i. e., the given sequence does not minimize the function L over E. Then there exists a number $\varepsilon > 0$ such that

$$\lim_{S \to \infty} L\big(\overline{y}_S(t),\, \overline{u}_S(t)\big) - l > \varepsilon,$$

and by (1.18)

$$\sup_E \int_{t_0}^{t_1} R(t,\, y(t),\, u(t))\, dt - \int_{t_0}^{t_1} \mu(t)\, dt > \varepsilon. \tag{1.19}$$

According to the definition of the exact upper bound, there exists a sequence $\{y_k(t),\, u_k(t)\} \subset E$ such that for a sufficiently large k

$$\sup_E \int_{t_0}^{t_1} R(t,\, y(t),\, u(t))\, dt - \int_{t_0}^{t_1} R(t,\, y_k(t),\, u_k(t))\, dt < \frac{\varepsilon}{2}. \tag{1.20}$$

Subtracting inequality (1.20) from (1.19), we find

$$\int_{t_0}^{t_1} R[t,\, y_k(t),\, u_k(t)]\, dt - \int_{t_0}^{t_1} \mu(t)\, dt > \frac{\varepsilon}{2},$$

or

$$\int_{t_0}^{t_1} \big[R(t,\, y_k(t),\, u_k(t)) - \sup_{(y,u)\in V(t)} R(t,\, y,\, u)\big]\, dt > \frac{\varepsilon}{2} > 0. \tag{1.21}$$

The integrand is non-positive almost everywhere in $(t_0,\, t_1)$, so that the integral is also non-positive, and inequality (1.19) breaks down.

The sequence $\{\overline{y}_S(t),\, \overline{u}_S(t)\}$ thus indeed minimizes the functional L over E.

In virtue of our lemma, this sequence also minimizes the functional I over D. Q. E. D.

§ 1.3.   PROBLEMS WITH A FREE BOUNDARY

So far we treated $[t_0,\, t_1]$ as a segment with fixed end points. In some problems, however, t_1 is not fixed and may be chosen from considerations of optimality.

We again consider the minimum of the functional

$$I = \int_{t_0}^{t_1} f^0(t,\, y,\, u)\, dt + F(y_0,\, y_1,\, t_1). \tag{1.22}$$

Here t_1 is an element of the set τ of points of the t axis contained in the segment $[t_0, T]$, $T < \infty$. The vector functions $y(t)$, $u(t)$ satisfy all the conditions listed in § 1.1.

The sets $V_y(t)$ and $V_u(t)$ are assumed to be defined over the segment $[t_0, T]$. The function $F(t_1, y_0, y_1)$ is defined and continuous for $t_1 \in \tau$, $y_0 \in V_y(t_0)$, $y_1 \in V_y(t_1)$. Each element of the set D is now a combination of a number $t_1 \in \tau$ and the vector functions $y(t)$ and $u(t)$ defined over the segment $[t_0, t_1]$.

The analog of Theorem 1.1 in this case is formulated as follows.

Theorem 1.2. Consider a sequence $\{\bar{t}_{1S}, \bar{y}_S(t), \bar{u}_S(t)\} \subset D$. For this sequence to minimize the functional I over D, it is sufficient that there exists a function $\varphi(t, y)$ (see § 1.2) such that

1) $\quad R(t, \bar{y}_S(t), \bar{u}_S(t)) \to \mu(t)$ \hfill (1.23)

on $(t_0, \bar{t}_{1S})$;

2) $\quad \Phi(\bar{t}_{1S}, \bar{y}_{0S}, \bar{y}_{1S}) \to \inf_{t_1 \in \tau, y_0 \in V_y(t_0), y_1 \in V_y(t_1)} \Phi(t_1, y_0, y_1) > -\infty;$ \hfill (1.24)

3) $\mu(t) \equiv 0$ on (t_0, T);

4) there exists a finite number Q such that for any $S < \infty$ and any $t \in (t_0, t_{1S})$:

$$R(t, \bar{y}_S(t), \bar{u}_S(t)) > Q,$$

where $R(t, y, u)$ and $\mu(t)$ are defined by (1.9), (1.10).

Proof. We define a set E of the triads $(t_1, y(t), u(t))$ which differs from D in two respects: first, the functions $y^i(t)$ $(i = 1, 2, \ldots, n)$ may have discontinuities of the first kind at a finite number of points of the segment $[t_0, T]$ and, second, the vector functions $y(t), u(t)$ are no longer related by differential equations (1.3). On E we define the functional

$$L(t_1, y(t), u(t)) = \Phi(t_1, y_0, y_1) - \int_{t_0}^{t_1} R(t, y, u)\,dt. \tag{1.25}$$

Clearly $L = I$ for $(t_1, y(t), u(t)) \in D$. We will now show that conditions 1 through 4 of the theorem are sufficient for the sequence $\{\bar{t}_{1S}, \bar{y}_S(t), \bar{u}_S(t)\}$ to minimize the functional L over E. In all other respects, the proof coincides with that to Theorem 1.1.

Let

$$k(t_1) = \inf_{(y(t), u(t)) \in E(t_1)} L(t_1, y(t), u(t)), \tag{1.26}$$

where $E(t_1)$ is the section of the set E for a fixed t_1, i.e., the set of elements $[t_1, y(t), u(t)] \in E$ for a fixed t_1,

$$\psi(t_1) = \inf_{(y_0, y_1) \in V_y(t_0) + V_y(t_1)} \Phi(t_1, y_0, y_1); \tag{1.27}$$

$$m = \inf_{t_1 \in \tau} \psi(t_1). \tag{1.28}$$

Using (1.26), (1.27), (1.28) and condition 3 of the theorem, we may write

$$I = \inf_E L(t_1, y(t), u(t)) = \inf_{t_1 \in \tau} k(t_1) = \inf_{t_1 \in \tau} \psi(t_1) = m. \tag{1.29}$$

On the other hand, the limit of the functional L over any sequence $\{t_{1S}, y_S(t), u_S(t)\} \in E$ satisfying the conditions of the theorem is also equal to m:

$$\lim_{S\to\infty} L\left(t_{1S},\ y_S(t),\ u_S(t)\right)=\lim_{S\to\infty}\left\{\Phi\left(t_{1S},\ y_{0S},\ y_{1S}\right)-\right.$$

$$\left.-\int_{t_0}^{t_{1S}} R\left(t,\ y_S(t),\ u_S(t)\right)dt\right\}=\lim_{S\to\infty}\Phi\left(t_{1S},\ y_{0S},\ y_{1S}\right)-$$

$$-\lim_{S\to\infty}\int_{t_0}^{t_{1S}} R\left(t,\ y_S(t),\ u_S(t),\ u_S(t)\right)dt. \tag{1.30}$$

Since the set τ is bounded, we see, using the properties of the sequence $R(t,\ y_S(t),\ u_S(t))$, that the second term vanishes. Therefore, using the last equality together with (1.24), (1.27), (1.28), we find

$$\lim_{S\to\infty} L\left(\bar{t}_{1S},\ \bar{y}_S(t),\ \bar{u}_S(t)\right)=m=l,$$

i. e., the sequence $\{t_{1S},\ \bar{y}_S(t),\ \bar{u}_S(t)\}$ indeed minimizes the functional L on E. The rest of the proof is conducted as for Theorem 1.1.

Chapter II

SOME METHODS OF SOLUTION OF VARIATIONAL PROBLEMS USING THE SUFFICIENT CONDITIONS OF THE ABSOLUTE MINIMUM

In this chapter we consider some methods of solution of variational problems based on Theorems 1.1 and 1.2; these are methods which show how to choose the function $\varphi(t, y)$ so as to reduce the treatment to standard classical methods — Lagrange's method and the Hamilton-Jacobi method — modifying them to such an extent that a complete solution of the problem is obtained, i. e., the absolute minimizing solution is found.

All the results are derived for the case when the minimizing solution $(\bar{y}(t), \bar{u}(t)) \in D$ exists. A more general case is treated in the next chapter.

§ 2.1. THE LAGRANGE-PONTRYAGIN METHOD

This method enjoys the widest applicability, although it is not always the simplest. The underlying idea is to find the partial derivatives of $\varphi(t, y)$ at the points of the minimizing solution

$$\psi_i(t) = \varphi_{y_i}[t, \bar{y}(t)], \ i = 1, 2, \ldots, n, \tag{2.1}$$

and also the minimizing solution $(\bar{y}(t), \bar{u}(t))$ itself, while the proof of the existence of the function $\varphi(t, y)$ satisfying the conditions of Theorem 1.1 is postponed to a later stage.

For simplicity, let t_1 be fixed, let $V_y(t)$ be an open region for all $t \in (t_0, t_1)$, which for $t = t_0$ and $t = t_1$ reduces to the points

$$y(t_0) \equiv y_0; \ y(t_1) \equiv y_1; \tag{2.2}$$

the set V_u is independent of y. In what follows, the functional I is conveniently replaced by a functional $\psi_0 I$, where ψ_0 is a positive constant. This substitution evidently does not affect the essential features of the problem.

We moreover assume that the functions $f^i(t, y, u)$, $i = 1, 2, \ldots, n$, are continuous and differentiable for all t, y, u. If we further demand that $\varphi(t, y)$ be twice continuously differentiable at the points corresponding to the presumably minimizing solution $\bar{y}(t)$, we can write the necessary conditions of a maximum of R in the form

$$R_{y^k}(t,\,\bar y,\,\bar u)=\varphi_{yy^k}f(t,\,\bar y,\,\bar u)+\varphi_y f_{y^k}(t,\,\bar y,\,\bar u)-$$
$$-\psi_0 f^0_{y^k}(t,\,\bar y,\,\bar u)+\varphi_{ty^k}=0;$$
$$k=1,\,2,\,\ldots,\,n$$
$$\varphi_y f(t,\,\bar y,\,\bar u)-\psi_0 f^0(t,\,\bar y,\,\bar u)=$$
$$=\sup_{u\in V_u}[\varphi_y f(t,\,\bar y,\,u)-f^0(t,\,\bar y,\,u)].\tag{2.3}$$

Using (2.1) and the equality $f(t,\,\bar y,\,\bar u)=\dot{\bar y}$, $(\bar y(t),\,\bar u(t))\in D$, we obtain

$$R_{y^k}[t,\,\bar y(t),\,\bar u(t)]\equiv\frac{d}{dt}\,\psi_k+H_{y^k}=0;\tag{2.4}$$

$$H[t,\,\psi(t),\,\bar y,\,\bar u]=\sup_{u\in V_u}H[t,\,\bar y(t),\,u],\tag{2.5}$$

where $\psi(t)=\dfrac{\partial\varphi(t,\,y)}{\partial y}\bigg|_{y=\bar y(t)}$ is the gradient of the function $\varphi(t,\,y)$ at a point $y=\bar y(t)$ in the space Y,

$$H(t,\,y,\,u)=\psi(t)f(t,\,y,\,u)-\psi_0 f^0(t,\,y,\,u).\tag{2.6}$$

Conditions (2.4) and (2.5), together with (1.3), constitute $2n+r$ equations in $2n+r$ unknown functions

$$y^i(t),\,\psi_i(t),\,u^s(t),\,i=1,\,2,\,\ldots,\,n;\,s=1,\,2,\,\ldots,\,r.$$

Together with boundary conditions (2.2), these equations define an e x t r e m a l, i.e., a pair $(\bar y(t),\,\bar u(t))\in D$ which satisfies the necessary conditions (2.3) for the supremum of $R(t,\,y,\,u)$, and a vector function $\psi(t)$ or, by (2.6), the gradient of the function $\varphi(t,\,y)$ at the points of the extremal.

Conditions (1.3) and (2.4) constitute a system of ordinary differential equations for the functions $\bar y^i(t)$, $\psi_i(t)$, $i=1,\,2,\,\ldots,\,n$, closed by the finite relation (2.5), according to which for $u=\bar u(t)$ the function $H[t,\,\bar y(t),\,\psi(t),\,u]$ reaches its largest value compared to its values for all the admissible control functions for every fixed $t\in(t_0,\,t_1)$. Equations (1.3), (2.4) may be written in the form

$$\dot y=\frac{\partial H}{\partial\psi};\tag{2.7}$$

$$\dot\psi=-\frac{\partial H}{\partial y}.\tag{2.8}$$

These equations constitute a so-called H a m i l t o n i a n s y s t e m and the function $H(t,\,y,\,\psi,\,u)$ is known as the H a m i l t o n f u n c t i o n or the H a m i l t o n i a n.

The variables y^i and ψ_i and the respective systems (2.7) and (2.8) are said to be c o n j u g a t e.

Equations (2.4), (2.5) prove to be not only the necessary conditions for a maximum of the function R, but also the necessary conditions for a minimum of the functional, provided the strict inequality $\psi_0>0$ is replaced by the weaker condition $\psi_0\geqslant0$. This is so because, after the inequality adjustment, these equations coincide with the necessary conditions for the minimum of the functional corresponding to Pontryagin's maximum principle.

At all points of the space $T \times Y$ which are not points of the extremal, the function $\varphi(t, y)$ may be defined quite arbitrarily. If there exists a function $\varphi(t, y)$ such that at the points of the extremal $\bar{y}(t)$, $\bar{u}(t)$ both the necessary and the sufficient conditions for the maximum of $R(t, y, u)$ are satisfied for every fixed $t \in (t_0, t_1)$, Theorem 1.1 indicates that the extremal $(\bar{y}(t), \bar{u}(t))$ is in fact the absolute minimizing solution. This result can be summarized by the following theorem.

T h e o r e m 2.1. Let the functions $\bar{y}(t)$, $\bar{u}(t)$, $\psi(t)$ constitute a solution of equations (1.3), (2.4), (2.5). A sufficient condition for the extremal $(\bar{y}(t), \bar{u}(t))$ to be an absolute minimizing solution for the functional (1.1) is the existence of a function $\varphi(t, y)$ which is continuously defined for all $t \in [t_0, t_1]$, $y \in V_y(t)$ and is twice differentiable with respect to y for all $y \in V_y(t)$, $t \in [t_0, t_1]$ and piecewise-differentiable with respect to t, such that

1) $\varphi_y{}^i [(t, \bar{y}(t)] = \psi_i(t), \ i = 1, 2, \ldots, n;$

2) $R(t, \bar{y}, \bar{u}) = \sup_{(y, u) \in V(t)} R(t, y, u), \ t \in (t_0, t_1)$

This theorem gives a sufficient condition of the absolute minimum which, unlike Weierstrass's sufficient condition of variational calculus and its generalizations /2/, does not require the construction of the field of extremals nor the derivation of any other extremals except the one being considered $\bar{y}(t)$, $\bar{u}(t)$.

The algorithm of the method reduces to solving a boundary-value problem for the system of ordinary differential equations (1.3), (2.4), (2.5) and choosing a function $\varphi(t, y)$ which satisfies the conditions of Theorem 2.1 or, more precisely, proving the existence of such a function. This is a highly significant qualification: it shows that in Lagrange's method we do not have to determine the particular form of the function $\varphi(t, y)$ satisfying the conditions of Theorem 2.1, and it is sufficient to prove the existence of this function only.

In particular, we may start with the expression

$$\varphi(t, y) = \psi_i(t) y^i + \sigma_{ij}(t) \Delta y^i \Delta y^j, \tag{2.9}$$

$$i, j = 1, 2, \ldots, n,$$

where

$$\Delta y^i = y^i - \bar{y}^i(t)$$

and the functions

$$\sigma_{ij}(t) = \varphi_{y^i y^j} [t, \bar{y}(t)] \tag{2.10}$$

are continuous and piecewise-differentiable over $[t_0, t_1]$. Theorem 2.1 then leads to Corollary 2.1.

C o r o l l a r y 2.1. For the extremal $\bar{y}(t)$, $\bar{u}(t)$ to be the absolute minimizing solution of the functional (1.1), it is sufficient that there exist n^2 functions $\sigma_{ij}(t)$ such that

$$R[t, \psi(t), \sigma(t), \bar{y}, \bar{u}] = \mu(t), \tag{2.11}$$

where

$$R = (\psi_i + \sigma_{ij} \cdot \Delta y^j) f^i(t,\ y,\ u) - f^0(t,\ y,\ u) +$$
$$+ \frac{d}{dt} \psi_i(t) y^i + \frac{\partial}{\partial t} [\sigma_{ij}(t) \Delta y^i \Delta y^j]. \tag{2.12}$$

E x a m p l e 2.1. Let us apply the above method to investigate the functional

$$\left. \begin{array}{l} I = \int\limits_0^{t_1} (u^2 - y^2)\, dt, \\[2ex] \dot{y} = u;\ \ y(0) = y(t_1) = 0. \end{array} \right\} \tag{2.13}$$

First let us find an extremal. We have

$$H = \psi u - u^2 + y^2.$$

Equations (2.7), (2.8) are written in the form

$$\dot{\psi} = -2y;\ \ \dot{y} = u;$$
$$H_{uu} = -2 < 0.$$

If $t_1 \neq \pi m$ $(m = 1,\ 2, \dots)$ the unique solution of this system satisfying the boundary conditions is

$$y(t) \equiv u(t) \equiv \psi(t) = 0,\ t \in (t_0,\ t_1). \tag{2.14}$$

Let

$$\varphi(t,\ y) = \sigma(t) y^2;\ \ \varphi_y \equiv 2\,\sigma y.$$

Then

$$R(t,\ y,\ u) \equiv \varphi_y u - u^2 + y^2 + \varphi_t = (1 + \dot{\sigma}) y^2 + 2\sigma y u - u^2. \tag{2.15}$$

For the extremal (2.14) to be an absolute minimizing solution, it is sufficient that there exists a continuous function $\sigma(t)$ such that the quadratic form $-R(t, y, u)$ be positive definite for every $t \in (0,\ t_1)$. The necessary and sufficient condition of this is

$$-R_{uu} \equiv 2 > 0;\ \ R_{yy} R_{uu} - R_{yu}^2 \equiv -4(1 + \sigma^2 + \dot{\sigma}) > 0.$$

The first inequality is satisfied identically, whereas the second imposes a constraint on $\sigma(t)$:

$$1 + \sigma^2 + \dot{\sigma} \leqslant 0;\ t \in (0,\ t_1). \tag{2.16}$$

Thus, for the functional (2.13) to attain an absolute minimum on the extremal (2.14), it is sufficient that there exists a function $\sigma(t)$ satisfying (2.16). For example, let $t_1 = 1/3$. Let $\sigma(t) = -2t$. It is readily seen that (2.16) is satisfied everywhere on the segment $[0,\ 1/3]$, so that on this segment (2.14) is an absolute minimizing solution.

Let us find the maximal value of t_1 for which the extremal (2.14) satisfies (2.16), i.e., let us find a continuous function $\sigma(t)$ which satisfies (2.16) over the maximal interval $(0, t_1)$. This function is obtained by solving the equation

$$1+\sigma^2+\dot\sigma=0,$$

and choosing a solution which is continuous over the maximal interval $(0, t_1)$. The general solution of this equation is

$$\sigma(t)=-\tan(t+c).$$

A particular solution which is continuous over the maximal interval is

$$\sigma=-\tan\left(t+\frac{\pi}{2}\right).$$

The continuity interval of this solution is $(0, \pi)$. Thus for any $t_1<\pi$, the extremal (2.14) is the absolute minimizing solution. As we know $/7/$, the point $t=\pi$ is the conjugate of the point $(0, 0)$ and by Jacobi's necessary condition there is no minimum on the extremal (2.14) for $t_1>\pi$.

Our sufficient conditions in this case thus coincide with the necessary conditions "apart from the point $t_1=\pi$".

2.1.1. A strong local minimum

We say that the functional I has a strong local minimum on the extremal $\bar y(t)$, $\bar u(t)$ if there exists $\varepsilon>0$ such that

$$I\,[\bar y(t), \bar u(t)]<I\,[y(t), u(t)] \tag{2.17}$$

for all the pairs $(y(t), u(t))\in D$, $|y(t)-\bar y(t)|<\varepsilon$, everywhere on $[t_0, t_1]$.

Let f^i and f^0 be twice differentiable functions for any $t\in[t_0, t_1]$; $(u, y)\in V(t)$ and the set $V_u(t)$ is an open bounded region.

We have the following sufficient conditions for a strong local minimum.

T h e o r e m 2.2. Let the vector functions $\bar y(t)$, $\bar u(t)$, $\psi(t)$ constitute a solution of system (1.3), (2.4), (2.5). Among these functions, $\bar u(t)$ is a strict maximum of $H(t, u)$ everywhere on $[t_0, t_1]$, i. e.,

$$H(t, u)<H(t, \bar u),\ u\in V_u(t),\ u\neq\bar u, \tag{2.18}$$

and the matrix $\|-H_{u^s u^t}\|$, $s,\ t=1, 2,\ldots,\ r$ is positive definite.

Then a sufficient condition for the functional I to have a strong local minimum on the extremal $(\bar y(t), \bar u(t))$ is the existence of n^2 continuous piecewise-differentiable functions $\sigma_{ij}(t)$ which on $[t_0, t_1]$ satisfy n differential inequalities

$$\Delta_{r+j}[t, \sigma(t), \dot\sigma(t)]>0, j=1, 2, \ldots, n. \tag{2.19}$$

Here

$$\Delta_{r+j}=\begin{vmatrix} -\bar R_{u^1u^1} & \cdots & -\bar R_{u^1u^r} & -\bar R_{u^1y^1} & \cdots & -\bar R_{u^1y^j} \\ -\bar R_{u^2u^1} & \cdots & -\bar R_{u^2u^r} & -\bar R_{u^2y^1} & \cdots & -\bar R_{u^2y^j} \\ -\bar R_{y^1u^1} & -\bar R_{y^1u^2} \cdots -\bar R_{y^1u^r} & -\bar R_{y^1y^1} & \cdots & -\bar R_{y^1y^j} \\ -\bar R_{y^ju^1} & -\bar R_{y^ju^2} & \cdots\cdots\cdots\cdots\cdots & -\bar R_{y^jy^j} \end{vmatrix} \tag{2.20}$$

$$\overline{R}_u{}^s{}_u{}^t = \overline{H}_u{}^s{}_u{}^t;$$
$$\overline{R}_u{}^s{}_{y^j} = \sigma_{ik}\overline{f}_u^k{}^s + \overline{H}_u{}^s{}_{y^i};$$
$$\overline{R}_{y^i y^j} = \dot{\sigma}_{ij} + \sigma_{ik}f_{y^j}^k + \sigma_{jk}f_{y^i}^k + H_{y^i y^j};$$
$$i,\,j,\,k = 1,\,2,\,\ldots,\,n;\ s,\,t = 1,\,2,\,\ldots,\,r. \tag{2.21}$$

(summation over k is implied).

The bar denotes the values of the corresponding functions for $y = \bar{y}(t)$, $u = \bar{u}(t)$.

Remark 2.1. The conditions of the theorem incorporate the conditions of negative definiteness of the quadratic form

$$d^2 R\,[t,\,y,\,\bar{u}] = \overline{R}_{y^i y^j}\Delta y^i \Delta y^j + 2\overline{R}_y{}^t{}_u{}^s \Delta y^i \Delta u^s +$$
$$+ R_u{}^s{}_u{}^t \Delta u^s \Delta u^t \tag{2.22}$$

for any fixed $t \in [t_0,\,t_1]$ (summation over repeating indices is implied).

Proof. The function $\varphi(t,\,y)$ is chosen in the form (2.9). Then $R(t,\,y,\,u)$ takes the form (2.12), where the matrix $\sigma_{ik}(t)$ satisfies inequalities (2.19).

By (2.1), there exists $\varepsilon_1 > 0$ such that

$$R(t,\,y,\,u) < R(t,\,\bar{y}(t),\,\bar{u}(t)) \tag{2.23}$$

for $|y - \bar{y}| < \varepsilon_1$, $|u - \bar{u}| < \varepsilon_1$ everywhere on $(t_0,\,t_1)$.

Now, by (2.18), we have

$$R(t,\,\bar{y},\,u) < R(t,\,\bar{y},\,\bar{u}),\ u \in V_u(t),\ u \neq \bar{u},$$

i.e., for all $u \in V_u(t)$, with the exception of $\bar{u}$.

Since $R(t,\,y,\,u)$ is continuous, for every fixed t there exists $\varepsilon_2 > 0$ such that

$$R(t,\,y,\,u) < R\left(t,\,\bar{y},\,\tilde{u}\right) \tag{2.24}$$

for

$$|y - \bar{y}| < \varepsilon_2;\ \tilde{u} \in V_u(t);\ |u - \tilde{u}| < \varepsilon_2;\ t \in (t_0,\,t_1).$$

Let ε be the smaller of the two numbers ε_1 and ε_2. Then, by (2.24),

$$R(t,\,y,\,u) < R(t,\,\bar{y},\,\bar{u});$$
$$|y - \bar{y}| < \varepsilon,\ u \in V_u(t),\ u \neq \bar{u},\ t \in (t_0,\,t_1) \tag{2.25}$$

and it follows from Theorem 1.1 that the functional I has a strong local minimum.

Remark 2.2. Theorem 2.2 is formulated assuming an open set V_u. If V_u is a closed region, we have the following analogous theorem.

Theorem 2.3. Let the vector functions $\bar{y}(t)$, $\bar{u}(t)$, $\psi(t)$ constitute a solution of systems (1.3), (2.4), (2.5). Here $\bar{u}(t)$ is a strict maximum of $H(t,\,\bar{y},\,\psi,\,u)$. Then a sufficient condition for a strong local maximum on the extremal $(\bar{y}(t),\,\bar{u}(t))$ is the existence of n^2 continuous piecewise-differentiable functions $\sigma_{ij}(t)$ and a number $\varepsilon > 0$ such that for all nonzero

$$|\Delta y^i| < \varepsilon, \quad |\Delta u^s| < \varepsilon$$
$$j,\ i = 1,\ 2,\ \ldots,\ n,\ s = 1,\ 2,\ \ldots,\ r;\ u \in V_u;\ t \in (t_0,\ t_1) \tag{2.26}$$

we have the inequality

$$\frac{1}{2}\, d^2R(t,\ \bar{y},\ u) + dR(t,\ \bar{y},\ \bar{u}) = \bar{R}_y \Delta y + R_u(t,\ \bar{y},\ \bar{u})\, \Delta u +$$
$$+ \frac{1}{2}\,(\bar{R}_{u^s u^t} \Delta u^s \Delta u^t + 2\bar{R}_{u^s y^i} \Delta u^s \Delta y^i + \bar{R}_{y^i y^j} \Delta y^i \Delta y^j) < 0. \tag{2.27}$$

Here R_{yy}, $R_{u^s u^t}$, $R_{u^s y}$ are the coefficients defined by (2.21),

$$\Delta y = y - \bar{y};$$
$$\Delta u = u - \bar{u}.$$

Suppose that a solution $\sigma_{ij}(t)$ of the system of differential inequalities (2.19) has been found and the existence of a strong local minimum on the extremal $(\bar{y}(t), \bar{u}(t))$ has thus been proved. By Corollary 2.1, the pair $(\bar{y}(t), \bar{u}(t))$ is the sought solution if the function $R(t,\ y,\ u)$ defined by (2.12) has an absolute, as well as a local, maximum on $V(t)$ at the point $(\bar{y}(t),\ \bar{u}(t))$ for any $t \in (t_0,\ t_1)$.

Note that the proof of the existence of an absolute minimum on the extremal $(\bar{y}(t),\ \bar{u}(t))$ is often conducted in a straightforward manner, without first proving the existence of a local minimum. This, in particular, is the approach used in the following two problems.

2.1.2. Systems linear in the phase coordinates

Let the right-hand sides of equations (1.3) and the integrand in (1.1) have the form

$$f^i = a^i_j(t)\, y^j + h^i(t,\ u); \quad i,\ j = 1,\ 2,\ \ldots,\ n; \tag{2.28}$$
$$f^0 = a^0_j(t)\, y^j + h^0(t,\ u). \tag{2.29}$$

The boundary conditions are fixed:

$$y_0 = y_{0f};\ y_1 = y_{1f}.$$

Equations (2.4) and (2.5) in this case take the form

$$\dot{\psi}_j + a^i_j \psi_i - a^0_j = 0; \tag{2.30}$$
$$H^*(t,\ \bar{u}) \equiv \psi_i(t) h^i(t,\ \bar{u}) - h^0(t,\ \bar{u}) = \sup_{u \in V_u(t)} H^*(t,\ u). \tag{2.31}$$

Suppose that $(\bar{y}(t),\ \bar{u}(t),\ \psi(t))$ is a solution of (2.28), (2.30), (2.31). We will show that the extremal $(\bar{y}(t),\ \bar{u}(t))$ is an absolute minimizing solution of the functional over the set D.

Let

$$\varphi(t, y) = \psi_i(t) y^i, \quad i = 1, 2, \ldots, n. \tag{2.32}$$

Then

$$R(t, y, u) = (\psi_i a_j^i + \dot{\psi}_j - a_j^0) y^j + H^*(t, u) = H^*(t, u). \tag{2.33}$$

Since $R(t, y, u)$ is independent of y, using (2.31) we find for all $t \in (t_0, t_1)$

$$R(t, \bar{y}, \bar{u}) = \sup_{(y, u) \in V(t)} R(t, y, u) = \mu(t). \tag{2.34}$$

Thus by Theorem 1.1 the pair $(\bar{y}(t), \bar{u}(t))$ is the absolute minimizing solution.

2.1.3. The problem of the minimal mean error

Let the right-hand sides of equations (1.3) have the form

$$f^i = a_j^i(t) y^j + b_s^i(t) u^s,$$
$$i, j = 1, 2, \ldots, n; \; s = 1, 2, \ldots, r, \tag{2.35}$$

and the function $f^0(t, y, u)$ is non-negative and convex on $V(t)$, $t \in (t_0, t_1)$. This is generally the final form to which the problem of optimal control of the perturbed state of a system near some given state $y(t) = 0$ is reduced. Equations (2.35) are thus the equations of the perturbed state, and the function (1.1) is a measure of deviation of the system from the given state.

Equations (2.4), (2.5) defining the extremal have the form

$$\dot{\psi}_j + a_j^i \psi_i - f_{y^j}^0 = 0; \tag{2.36}$$

$$H(t, \bar{y}(t), \psi(t), \bar{u}) = \sup_{u \in V_u(t)} H \cdot (t, \bar{y}(t), \psi(t), u). \tag{2.37}$$

Suppose that $(\bar{y}(t), \bar{u}(t), \psi(t))$ is a solution of system $(2.35) - (2.37)$. We will show that the extremal $(\bar{y}(t), \bar{u}(t))$ is an absolute minimizing solution. We take $\varphi(t, y)$ in the form (2.32). Then

$$R(t, y, u) = (\psi_i a_j^i + \dot{\psi}_j) y^j + \psi_i(t) b_s^i u^s - f^0(t, y, u).$$

The function $R(t, y, u)$ for every fixed t is convex and it thus has a single supremum $\bar{y}, \bar{u}$ on $V(t)$. Therefore, by Theorem 1.1, the pair $(\bar{y}(t), \bar{u}(t))$ is an absolute minimizing solution.

2.1.4. Jacobi's necessary and sufficient condition of the variational calculus

Using the particular case of a very simple functional, we will establish a relationship between the conditions of a maximum of the function R and Jacobi's variational condition of a weak local minimum of a functional.

Consider the problem of minimizing the functional

$$I=\int_{t_0}^{t_1} f^0(t,\ y,\ u)\ dt;\qquad\qquad (2.38)$$

$$\dot{y}=u,$$

$$y_0,\ y_1\ \text{are given,}$$

where $y(t)$, $u(t)$ are scalar functions, $f^0(t,\ y,\ u)$ is a continuous function with continuous partial derivatives to third order inclusive.

Let $V(t)$ with $t\in(t_0,\ t_1)$ be an open domain.

Let the functions $\varphi(t,\ y)$ have continuous partial derivatives with respect to t and y to third order inclusive.

The function R is written in the form

$$R(t,\ y,\ u)=\varphi_y u - f^0(t,\ y,\ u)+\varphi_t(t,\ y).\qquad (2.39)$$

For sup R with respect to y and u to be attained on the pair $(\bar{y}(t),\ \bar{u}(t))\in D$, it is necessary that

$$\left.\begin{aligned}
\overline{R}_y&=\dot{\psi}+\overline{H}_y\equiv\dot{\psi}-\overline{f}_y^0=0;\\
\overline{R}_u&\equiv\overline{H}_u\equiv\psi-\overline{f}_u^0=0.
\end{aligned}\right\}\qquad (2.40)$$

Note that conditions (2.40), together with equation (2.38), are equivalent to Euler's equation (which may be obtained by eliminating ψ and u between the appropriate equations).

The next necessary condition of a maximum of R is that the quadratic form

$$d^2R=\overline{R}_{yy}\Delta y^2+2\overline{R}_{yu}\Delta y\Delta u+\overline{R}_{uu}\Delta u^2\qquad (2.41)$$

be non-positive; here bars denote the values of the second-order derivatives corresponding to the pair $(\bar{y}(t),\ \bar{u}(t))$.

Setting $\sigma(t)=\varphi_{yy}(t,\ \bar{y}(t))$, we obtain

$$\overline{R}_{uu}=-f_{uu}^0\left(t,\ \bar{y},\ \bar{u}\right)=-\overline{f}_{uu}^0;\qquad (2.42)$$

$$\overline{R}_{yu}=\sigma-\overline{f}_{yu}^0;\qquad (2.43)$$

$$\overline{R}_{yy}=\dot{\sigma}-\overline{f}_{yy}^0.\qquad (2.44)$$

Let $\bar{y}(t)$ be a continuous function. Then $\sigma(t)$ is also continuous.

A quadratic form is non-positive if and only if the diagonal minors of the matrix

$$\left\|\begin{matrix}-\overline{R}_{yy} & -\overline{R}_{yu}\\ -\overline{R}_{yu} & -\overline{R}_{uu}\end{matrix}\right\|$$

are non-negative /4/. Using (2.42), we can thus write

1) $\overline{f}_{uu}^0\geqslant 0;$ $\qquad\qquad (2.45)$

2) $-\overline{f}_{uu}^0(\dot{\sigma}-\overline{f}_{yy}^0)-(\sigma-\overline{f}_{yu}^0)^2\geqslant 0.$ $\qquad\qquad (2.46)$

The first inequality in (2.45) is Legendre's well-known condition. Let us consider in some detail condition (2.46): this is a differential inequality involving the function $\sigma(t)$. Let $\bar{f}^0_{uu} > 0$ everywhere in $[t_0, t_1]$. Note that since $\sigma(t)$ is quite arbitrary, condition (2.46) can always be satisfied on a sufficiently small interval (t_0, t_1) by an appropriate choice of the initial conditions $\sigma(t_0)$ and $\sigma(t_0)$. To check whether or not (2.46) is satisfied on the entire interval (t_0, t_1), we write it in the form of a Riccati equation for the function $\sigma(t)$:

$$-\bar{f}^0_{uu}\left(\dot{\sigma} - \bar{f}^0_{yy}\right) - \left(\sigma - \bar{f}^0_{yu}\right)^2 = \varepsilon(t)\bar{f}^0_{uu}, \qquad (2.47)$$

where $\varepsilon(t) \geqslant 0$ is some continuous function of t. We know from the theory of differential equations that the singular points of the solution of (2.47) (the points where $\sigma(t)$ does not exist) coincide with the zeros of the non-trivial solution of the second-order homogeneous linear equation

$$\frac{d}{dt}\left(\bar{f}^0_{uu}\dot{v}\right) + \left[\frac{d}{dt}\bar{f}^0_{yu} - \bar{f}^0_{yy} + \varepsilon\right] v = 0, \qquad (2.48)$$

which is obtained from (2.47) by a transformation

$$\sigma - \bar{f}^0_{yu} = \frac{\dot{v}}{v}\,\bar{f}^0_{uu}. \qquad (2.49)$$

For $\varepsilon = 0$ equation (2.48) reduces to Jacobi's equation /2/

$$\frac{d}{dt}\left(\bar{f}^0_{uu}\dot{v}\right) - \left(\bar{f}^0_{yy} - \frac{d}{dt}\bar{f}^0_{yu}\right) v = 0. \qquad (2.50)$$

Equation (2.48) will be investigated using Sturm's theorem and the theorem of the alternating zeros of solutions of second-order linear homogeneous equations. The theorem of the alternating zeros indicates that the maximal interval (t_0, t_1) on which the solution $v(t)$ of equation (2.48) does not vanish for a given $\varepsilon(t)$ corresponds to the solution which satisfies the initial condition $v(t_0) = 0$; in other words, it corresponds to a solution with one of the zeros coinciding with the left-hand end point of (t_0, t_1). Let this solution be designated $v(t, t_0)$. Sturm's theorem leads to the following proposition.

Let $v(t_0) = 0$ for equation (2.48) for any $\varepsilon(t) \geqslant 0$. Then among all the equations (2.48) with any $\varepsilon \geqslant 0$, the maximal interval (t_0, t) on which the solution $v(t, t_0)$ does not vanish corresponds to Jacobi's equation.

In other words, under the above donditions, for $\varepsilon(t) \geqslant 0$, the next zero of the solution of (2.48) on the right of t_0 is no farther from t_0 than the corresponding zero of the solution of Jacobi's equation, generally called the conjugate point of $t = t_0$.

Thus, if $\bar{f}^0_{uu} > 0$ and there exists a continuous and differentiable function $\sigma(t)$ such that condition (2.46) is satisfied on (t_0, t_1), the interval (t_0, t_1) contains no points conjugate to 0, this being Jacobi's necessary condition for a minimum of the functional (2.38).

Conversely, the comparison theorem shows that if (t_0, t_1) does not contain points conjugate to t_0, there exists a function $\varepsilon(t) \geqslant 0$ such that the solution $v(t, t_0)$ of equation (2.48) does not vanish anywhere in (t_0, t_1).

If t_1 is the conjugate of t_0, we have $\varepsilon(t) \equiv 0$. Then there exists a corresponding solution of Riccati's equation or, in other words, a function $\sigma(t)$ such that (2.46) is satisfied.

Let us consider the conditions of negative definiteness of the form (2.41), namely that everywhere in $[t_0,\ t_1]$, $\bar{f}^0_{uu} > 0$, and on $(t_0,\ t_1)$ there exists a function $\sigma(t)$ such that the strengthened condition (2.46) (the strict inequality) is satisfied or, equivalently, $\varepsilon(t) > 0$ on $(t_0,\ t_1)$ in (2.47) and (2.48). The form (2.41) is then negative definite. We will show that when these conditions are satisfied, the segment $[t_0,\ t_1]$ does not contain points which are conjugate to t_0.

Suppose that this is not so, i. e., t_1 is the conjugate of t_0 (the interval $(t_0,\ t_1)$, as we have seen before, contains no conjugate points) and $v(t,\ t_0)$ vanishes at t_1. Then for any $\varepsilon(t) > 0$ the nearest zero of the solution of (2.48) is nearer than t_1, lying inside the interval $(t_0,\ t_1)$. This signifies that the quadratic form d^2R cannot be negative definite, at variance with the conditions of the theorem.

Conversely, let $\bar{f}^0_{uu} > 0$ on $[t_0,\ t_1]$ and let $[t_0,\ t_1]$ contain no points which are conjugate to t_0. Since the solution of Jacobi's equation for $v(t_0) = 0$ does not vanish on the half-open interval $[t_0,\ t_1]$, there exists $\delta > 0$ such that the solution of equation (2.50) does not vanish on $(t_0,\ t_1 + \delta)$ either.

Because of the continuous dependence of the solutions of linear differential equations on parameters, there exists a constant $\varepsilon > 0$ such that the solution of equation (2.48) does not vanish on $[t_0,\ t_1]$ either. Hence it follows that a function $\sigma(t)$ satisfying the strengthened condition (2.46) exists on $(t_0,\ t_1)$.

We have thus proved the following theorem.

T h e o r e m 2.4. Consider a pair $(\bar{y}(t),\ \bar{u}(t)) \in D$, where $\bar{u}(t)$ is continuous and $\bar{f}^0_{uu} > 0$ everywhere in $[t_0,\ t_1]$. A necessary and sufficient condition for the existence of a continuous and continuously differentiable function $\sigma(t)$ such that for any fixed $t \in (t_0,\ t_1)$ the quadratic form

$$d^2R = \bar{R}_{uu}\Delta u^2 + 2\bar{R}_{yu}\Delta y \Delta u + \bar{R}_{yy}\Delta y^2$$

is negative definite is that for the pair $\bar{y}(t),\ \bar{u}(t)$ the interval $(t_0,\ t_1)$ (or the segment $[t_0,\ t_1]$) contains no points conjugate to t_0.

Thus for a nondegenerate $(\bar{f}^0_{uu} \neq 0)$ classical Lagrange's problem, the condition of a non-positive quadratic form

$$\left.\begin{array}{l} d^2R \leqslant 0; \\ t \in (t_0,\ t_1) \end{array}\right\} \tag{2.51}$$

is equivalent to Jacobi's necessary condition of a weak local minimum, and the strengthened condition (2.51) is equivalent to Jacobi's strengthened condition. However, the region of application of condition (2.51) or of its extended version (2.27) is much wider than of Jacobi's condition: they are applicable to degenerate problems (when the determinant vanishes, $|H_{u^i u^j}(t,\ \bar{y},\ \bar{u})| = 0$) and to problems with a closed region V_u, where

Jacobi's condition does not apply. Note that it is for these problems that the analysis of the second-order conditions is especially important. They are needed not only for checking purposes, but are actually used constructively, since they provide a means for isolating the local minimizing solution from the set of all extremals (in these cases, as distinct from the case of nondegenerate problems, the extremals are not unique).

Example 2.2. As an example of such a problem, let us minimize the functional

$$I = - \int\limits_{-1}^{+1} ty^2 dt;$$
$$\dot{y} = u; \quad |u| \leqslant 1;$$
$$y(-1) = y(1) = 0. \qquad (2.52)$$

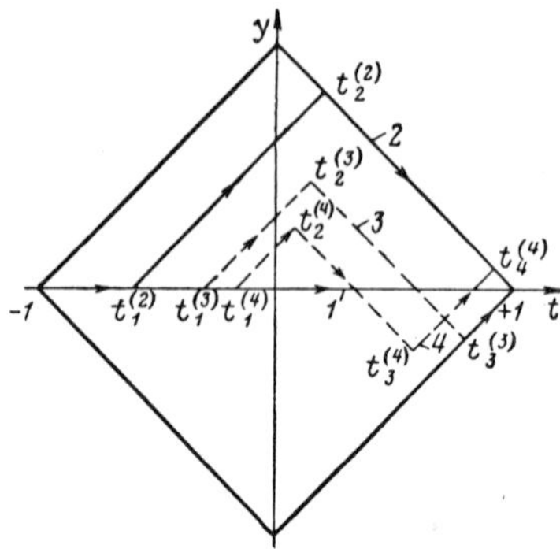

FIGURE 2.1

Conditions (2.4), (2.5) take the form

$$\dot{\psi} = - 2ty; \qquad (2.53)$$

$$\overline{H} = \psi\overline{u} + t\overline{y}^2 = \sup_{|u| \leqslant 1} H(t, \overline{y}, u). \qquad (2.54)$$

The last condition corresponds to three types of control:

$$u = \begin{cases} +1 & \text{for} \quad \psi > 0 \\ -1 & \text{for} \quad \psi < 0 \\ \text{arbitrary within the limits } |u| \leqslant 1 & \text{for} \quad \psi \equiv 0. \end{cases} \qquad (2.55)$$

It is readily seen that conditions (2.52), (2.53), (2.55) are satisfied by an infinite set of extremals (Figure 2.1), which differ from one another in the number of control switching points and the number of isolated zeros of the function $\psi(t)$. The initial segment of each extremal corresponds to the third control type in (2.55). Indeed, differentiating the identity $\psi(t) \equiv 0$ and inserting (2.53), we find $\overline{y}(t) \equiv 0$.

To select the absolute minimizing solution from among these extremals, a more detailed investigation of the maximum of $R(t, y, u)$ is required.

We have

$$R = \varphi_y(t, y)u + ty^2 + \varphi_t(t, y).\qquad(2.56)$$

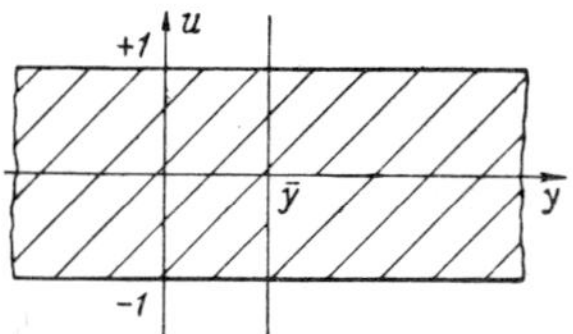

FIGURE 2.2

Here $\varphi(t, y)$ is regarded as a thrice differentiable function.

We seek a solution of the problem satisfying the condition of the fundamental theorem. Such a solution $\bar{y}(t)$, $\bar{u}(t)$ should satisfy the condition

$$R(t, \bar{y}, \bar{u}) \geqslant R(t, y, u)\qquad(2.57)$$

for all $(y, u) \in G$.

Here G is a horizontal strip on the y, u plane between the straight lines $u = \pm 1$ (Figure 2.2). For (2.57) to be satisfied on G, it is necessary in particular for (2.57) to be satisfied on a thin strip G_ε enclosing the segment $[-1, +1]$ of the vertical line $y = \bar{y}$ in the (y, u) plane. For those t when $\psi \neq 0$, we have $u = \pm 1$. The necessary condition for this is

$$R_{yy} \equiv \varphi_{yyy}(t, y)u + \varphi_{yyt} + 2t \Big|_{\substack{y = \bar{y}(t) \\ u = \bar{u}(t)}} \leqslant 0,$$

or, seeing that $\dot{\bar{y}} = \bar{u}$, and writing

$$\sigma(t) = \varphi_{yy}[t, \bar{y}(t)],$$

we obtain

$$R_{yy}(t, \bar{y}, \bar{u}) = \dot{\sigma} + 2t \leqslant 0.\qquad(2.58)$$

In G_ε we need only retain the first terms in the expansion of φ and R in powers of $\Delta y = y - \bar{y}$, and we may thus write

$$\varphi(t, y) = \varphi(t, \bar{y}(t)) + \psi(t)\Delta y + \frac{\sigma(t)}{2}\Delta y^2 + o(\Delta y^2);$$

$$R = (\psi + \sigma\Delta y)u + t(\bar{y} + \Delta y)^2 + \dot{\psi}(\Delta y) +$$

$$+ \frac{\dot{\sigma}}{2}(\Delta y)^2 + \dot{\varphi}(t, \bar{y}(t)) - (\psi(t) + \sigma(t)\Delta y)\dot{\bar{y}}.$$

We see from the last formula that for $\psi=0$ the function $R(y, u)$ has a supremum on G_ε at the points $(\bar{y}, 1)$ and $(\bar{y}, -1)$ for $\sigma=0$ only. Thus, for condition (2.57) to be satisfied for $t \in (-1, +1)$, it is necessary that there exists a continuous function $\sigma(t)$ satisfying (2.58) which vanishes for those t when $\psi=0$, i. e., at the switching points. It follows from (2.58) that for $t>0$, σ increases and consequently there may exist only one $t=\tau$, for which $\sigma=0$, i. e., (2.57) is satisfied only by an extremal containing at most one switching point with a positive abscissa. The only such extremal is extremal 2 in Figure 2.1.

Thus, the necessary condition for the existence of $\sigma(t)$ for which R has a maximum in y in a small neighborhood selects a single extremal — extremal 2 — from among an infinite number of extremals. This extremal is readily shown to satisfy the conditions of Theorem 1.1. Let φ be given in the form

$$\varphi(t, y) = \psi(t) y + \frac{\sigma(t)}{2} (y - \bar{y}(t))^2, \tag{2.59}$$

where $\sigma(t)$ is a continuous function satisfying inequality (2.58) and the condition $\sigma(t_2^{(2)}) = 0$, $\sigma(t) \equiv 0$ for $t < t_1^{(2)}$. Then R is a quadratic form and satisfies (2.57) for any t, whence it follows that extremal 2 is an absolute minimizing solution.

2.1.5. General boundary conditions

1. Let the sets $V_y(t_0)$ and $V_y(t_1)$ be some general sets in the space Y, and not the points y_0 and y_1 as assumed before; t_1 may take values from a given bounded set τ. The function $F(t_1, y_1, y_0)$ entering (1.1) is continuous and differentiable. In this case, additional conditions (1.24) of Theorems $2.1-2.4$ are imposed on the sought minimizing solution $\bar{y}(t)$, $\bar{u}(t)$ and the function $\varphi(t, y)$, which demand that the function $\Phi(y_0, y_1, t_1)$ have a minimum for $y_0 = \bar{y}_0$, $y_1 = \bar{y}_1$, $t_1 = \bar{t}_1$ and that $\mu(t) = 0$.

A necessary condition for a minimum of $\Phi(t_1, y_1, y_0)$ at the point $\bar{t}_1, \bar{y}_0, \bar{y}_1$ is provided by the inequality

$$\frac{\partial \overline{\Phi}}{\partial y_0} \Delta y_0 + \frac{\partial \overline{\Phi}}{\partial y_1} \Delta y_1 + \frac{\partial \overline{\Phi}}{\partial t_1} \Delta t_1 \geqslant 0 \tag{2.60}$$

for all

$$t_1 \in \tau, \quad y_0 \in V_y(t_0), \quad y_1 \in V_y(t_1),$$

where $\Delta y_0 = y_0 - \bar{y}_0$; $\Delta y_1 = y_1 - \bar{y}_1$; $\Delta t_1 = t_1 - \bar{t}_1$ are all sufficiently small (the bar denotes the values of the derivatives for $y_i = \bar{y}_i$, $t_1 = \bar{t}_1$).

Using (1.8), we write (2.60) in a more explicit form:

$$\left[\frac{\partial \overline{F}}{\partial y_0} - \varphi_y(t_0, \bar{y}_0) \right] \Delta y_0 + \left[\frac{\partial \overline{F}}{\partial y_1} + \varphi_y(t_1, \bar{y}_1) \right] \Delta y_1 +$$

$$+ \left[\frac{\partial \overline{F}}{\partial t_1} + \varphi_t(t_1, \bar{y}_1) \right] \Delta t_1 \geqslant 0.$$

Using the notation from (2.1) and (2.6), and seeing that in virtue of the identity $\mu(t) \equiv 0$

$$\frac{\partial \varphi(t, \overline{y}(t))}{\partial t} = f^0 - \frac{\partial \varphi(t, \overline{y}(t))}{\partial y} f(t, \overline{y}, \overline{u}) =$$

$$= -H(t, \overline{y}(t), \psi(t), \overline{u}(t)),$$

we finally obtain

$$\left[\frac{\partial \overline{F}}{\partial y_0} - \psi(t_0)\right] \Delta y_0 + \left[\frac{\partial \overline{F}}{\partial y_1} + \psi(\overline{t}_1)\right] \Delta y_1 +$$

$$+ \left[\frac{\partial \overline{F}}{\partial t_1} - H(\overline{t}_1, \overline{y}_1, \overline{u}(t_1))\right] \Delta t_1 \geqslant 0$$

for all

$$t_1 \in \tau; \quad y_1 \in V_y(t_1); \quad y_0 \in V_y(t_0). \tag{2.61}$$

This inequality, together with the conditions $y_0 \in V_y(t_0)$, $y_1 \in V_y(t_1)$ and $t \in \tau$, specifies $2n+1$ boundary conditions for the system of differential equations (1.3), (2.4), (2.5) and a finite value of the argument t_1.

These boundary conditions can be written in a more concrete form if the particular sets τ, $V_y(t_1)$, and $V_y(t_0)$ are defined explicitly.

In our previous example, when the sets were identified with fixed points in space, inequality (2.60) did not impose any constraints on the vectors $\psi(t_0)$, $\psi(t_1)$, since

$$\Delta y_0 = \Delta y_1 = \Delta t_1 = 0.$$

Another common type of boundary conditions corresponds to the following case: y_0 is a fixed point, the set $V_y(t_1)$ coincides with the entire space Y, i.e., there are no constraints on y_1, and the set τ corresponds to the semi-infinite interval (t_0, ∞). In this case, for inequality (2.60) to be satisfied, the coefficients before Δy_1 and Δt_1 must vanish, i.e.,

$$\psi(t_1) = -\frac{\partial \overline{F}}{\partial y_1}; \tag{2.62}$$

$$H(t_1, \overline{y}(t_1), \overline{u}(t_1)) = \frac{\partial \overline{F}}{\partial t_1}. \tag{2.63}$$

These equalities define the $n+1$ missing boundary conditions for the system (1.3), (2.4) and the time t_1.

When investigating the sufficient conditions for a minimum, the condition of a minimum of $\Phi(y_0, y_1, t_1)$ imposes additional boundary conditions on the differential inequalities for the functions $\sigma_{ij}(t)$. They may be derived from the inequality

$$d\Phi + \frac{1}{2} d^2\Phi = \frac{\partial \overline{\Phi}}{\partial y_0} \Delta y_0 + \frac{\partial \overline{\Phi}}{\partial y_1} \Delta y_1 + \frac{\partial \overline{\Phi}}{\partial t_1} \Delta t_1 +$$

$$+ \frac{1}{2} \left(\frac{\partial^2 \overline{\Phi}}{\partial y_0^i \partial y_0^j} \Delta y_0^i \Delta y_0^j + 2 \frac{\partial^2 \overline{\Phi}}{\partial y_0^i \partial y_1^j} \Delta y_0^i \Delta y_1^j + \right.$$

$$+ 2 \frac{\partial \bar{\Phi}}{\partial y_0^i \partial t_1} \Delta y_0^i \Delta t_1 + 2 \frac{\partial \bar{\Phi}}{\partial y_1^i \partial t_1} \Delta y_1^i \Delta t_1 +$$

$$+ \frac{\partial^2 \bar{\Phi}}{\partial y_1^i \partial y_1^j} \Delta y_1^i \Delta y_1^j + \frac{\partial^2 \bar{\Phi}}{(\partial t_1)^2} (\Delta t_1)^2 \Big) > 0, \qquad (2.64)$$

which should be satisfied for all sufficiently small non-zero increments Δy_0, Δy_1, Δt_1 when

$$t_1 \in \tau; \quad y_0 \in V_y(t_0); \quad y_1 \in V_y(t_1).$$

Here

$$\frac{\partial \bar{\Phi}}{\partial y_0} = - \psi(t_0) + \frac{\partial \bar{F}}{\partial y_0};$$

$$\frac{\partial \bar{\Phi}}{\partial y_1} = + \psi(t_1) + \frac{\partial \bar{F}}{\partial y_1};$$

$$\frac{\partial \bar{\Phi}}{\partial t_1} = \frac{\partial \bar{F}}{\partial t_1} - H(t, \bar{y}(t_1), \bar{u}(t_1));$$

$$\frac{\partial^2 \bar{\Phi}}{\partial y_0^i \partial y_0^j} = \frac{\partial^2 \bar{F}}{\partial y_0^i \partial y_0^j} - \sigma_{ij}(t_0);$$

$$\frac{\partial^2 \bar{\Phi}}{\partial y_0^i \partial y_1^j} = \frac{\partial^2 \bar{F}}{\partial y_0^i \partial y_1^j};$$

$$\frac{\partial^2 \bar{\Phi}}{\partial y_0^i \partial t_1} = \frac{\partial^2 \bar{F}}{\partial y_0^i \partial t_1};$$

$$\frac{\partial^2 \bar{\Phi}}{\partial y_1^i \partial t_1} = \frac{\partial^2 \bar{F}}{\partial y_1^i \partial t_1} + \dot{\psi}_i(t_1) - \sigma_{ij}(t_1) \dot{\bar{y}}^j(t_1);$$

$$\frac{\partial^2 \bar{\Phi}}{\partial y_1^i \partial y_1^j} = \frac{\partial^2 \bar{F}}{\partial y_1^i \partial y_1^j} + \sigma_{ij}(t_1);$$

$$\frac{\partial^2 \bar{\Phi}}{(\partial t_1)^2} = \frac{\partial^2 \bar{F}}{(\partial t_1)^2} - \dot{\bar{H}} - \dot{\psi}\dot{\bar{y}} + \sigma_{ij}\dot{\bar{y}}^i\dot{\bar{y}}^j.$$

Summation over repeating indices is assumed throughout.

For the previously considered particular case — the problem with a free right-hand end point — inequality (2.64) is equivalent to the condition of positive definiteness of the quadratic form

$$\frac{\partial^2 \bar{\Phi}}{\partial y_1^i \partial y_1^j} \Delta y_1^i \Delta y_1^j + 2 \frac{\partial^2 \bar{\Phi}}{\partial y_1^i \partial t_1} \Delta y_1^i \Delta t_1 + \frac{\partial^2 \bar{\Phi}}{(\partial t_1)^2} (\Delta t_1)^2.$$

2.1.6. Numerical aspects

Lagrange's method thus reduces the variational problem to finding the extremal $(\bar{y}(t), \bar{u}(t))$ and a vector function $\psi(t)$, i.e., to solving a system of ordinary differential equations of order $2n$ with boundary conditions (2.61), followed by verifying the sufficient conditions of optimality on this extremal, e.g., by solving the differential inequalities (2.19) for the functions $\sigma_{ij}(t)$.

It is significant that the boundary conditions are separated, i.e., some of them correspond to the value t_0 of the argument, while the remaining boundary conditions correspond to t_1. The result is a boundary-value problem, as distinct from the Cauchy problem, when all the boundary conditions are specified for the same value of the argument, t_0 or t_1. The majority of the modern computation algorithms solve a given boundary-value problem by constructing a sequence of solutions of Cauchy problems obtained by varying the missing initial conditions until the required final conditions are satisfied.

Let us consider the actual procedure for solving the Cauchy problem for system (1.3), (2.4) by Euler's polygonal method. According to this method, the segment $[t_0, t_1]$ is divided into s parts by the points $\tau_1 = t_0,\ \tau_2 > \tau_1,\ \tau_3 > \tau_2,\ \dots\ \tau_s = t_1$. Given $y(t_0)$, $\psi(t_0)$, we find $\bar{u}(t_0)$ from the condition of the maximum of the function $H(t, y(t_0), \psi(t_0), u)$ and, inserting the result on the right in (1.3), (2.4), we obtain $\dot{y}(t_0)$ and $\dot{\psi}(t_0)$. The finite-difference formulas of Euler's method

$$y(\tau_2) = y(t_0) + \dot{y}(t_0)(\tau_2 - \tau_1);$$

$$\psi(\tau_2) = \psi(t_0) + \dot{\psi}(t_0)(\tau_2 - \tau_1)$$

are then applied to determine $y(\tau_2)$ and $\psi(\tau_2)$. Reiterating the procedure, we find

$$y(\tau_3),\ \psi(\tau_3),\ \dots,\ y(t_1),\ \psi(t_1).$$

In solving the boundary-value problem, this procedure is repeated for various values of $y(t_0)$, $\psi(t_0)$ until $y(t_1)$, $\psi(t_1)$, t_1 satisfy the final conditions with sufficient accuracy.

This explains why the solution of boundary-value problems is so time consuming, even with modern computers.

Various methods exist which cut down the search for the initial values $y(t_0)$, $\psi(t_0)$ and often markedly reduce the volume of computation. These improved methods include Newton's method /3/, the method of fastest descent /9/, random access methods /5/. However, even with these methods, the solution of boundary-value problems involves significant difficulties, especially as none of them is absolutely reliable.

It should be emphasized that despite all the efforts, the extremal obtained in this way may prove not to be the sought minimizing solution, since equations (1.3), (2.4), (2.5) only give the necessary conditions for a maximum of R.

No adequate algorithms are available so far which would enable us to tackle the second part of the problem, namely the verification of the sufficient conditions. In a number of cases, this problem is solved relatively simply by choosing an appropriate function φ, as in 2.1.2 and 2.1.3 above. In other cases, the solution can be simplified by using inequalities (2.19) for $\sigma_{ij}(t)$.

§ 2.2. HAMILTON–JACOBI–BELLMAN METHOD. OPTIMAL CONTROL SYNTHESIS

In the present treatment, we choose a fixed t_1 and make $V_y(t)$ coincide with the entire space Y for $t \in (t_0, t_1]$ and with the point $y = y_0$ for $t = t_0$; $F(y_1, y_0)$ is independent of y_0.

We construct the function

$$P(t, y) = \sup_{u \in V_u(t, y)} R(t, y, u) \tag{2.65}$$

and try to select $\varphi(t, y)$ so that P is independent of y, i.e.,

$$P(t, y) \equiv \sup_{u \in V_u(t, y)} [\varphi_y f(t, y, u) - f^0(t, y, u)] + \varphi_t = C(t), \tag{2.66}$$

where $C(t)$ is an arbitrary piecewise-continuous function.

Then $P(t, y) = \mu(t)$ at any point of the space Y. Let $\bar{u}(t, y)$ be the value of u for which $R(t, y, u)$ attains a supremum at the point t, y, i.e.,

$$R(t, y, \bar{u}) = P(t, y) \tag{2.67}$$

and let $\bar{y}(t)$ be the solution of the system

$$\dot{y} = f[t, y, \bar{u}(t, y)] \tag{2.68}$$

with the boundary condition $y(0) = y_0$; also let $\bar{u}(t) = \bar{u}(t, \bar{y}(t))$. The pair of vector functions $(\bar{y}(t), \bar{u}(t))$ belongs to the class D and satisfies condition 1 of Theorem 1.1. For this pair to satisfy condition 2 of this theorem, i.e., to be an absolute minimizing solution, it suffices to take

$$F(y) + \varphi(t_1, y) = \text{const}, \tag{2.69}$$

i.e., it is sufficient that for $t = t_1$ the function $F + \varphi$ be independent of y. Thus, if the function $\varphi(t, y)$ can be selected so that $\sup_u R$ is independent of y, or more precisely, if we can solve the partial differential equation (2.66) with the boundary condition (2.69), our problem is completely solved. Moreover, a much more general problem is solved, namely how, starting from any given state $(t_0, y_0) \in V$, to reach the least value of the functional (1.1) in a time t_1. To accomplish this, at every point (t, y), starting with (t_0, y_0), we choose a control function $\bar{u}(t, y)$. If $\varphi(t, y)$ is known, $\bar{u}(t, y)$ is obtained from (2.66). The function $\bar{y}(t)$ is then found from (2.68).

Using the terminology of the automatic control theory, we refer to the solution of the variational problem constructed in this way as the optimal control synthesis, and to the field $\bar{u}(t, y)$ as the synthesizing function. In distinction from synthesis, the optimal control function $\bar{u}(t)$ corresponding to the given initial conditions (t_0, y_0) is called the optimal control program.

These two forms of solution of the problem correspond to two fundamentally different methods of practical optimal control. The synthesis solution is implemented by a closed automatic control system (a feedback system). In this system, the synthesizing function is regarded as an algorithm which determines the response of the feedback operator to the signals from the sensor elements measuring the time and the current state of the controlled object (Figure 2.3).

The control program solution is implemented by an open system, without feedback, as shown in Figure 2.4.

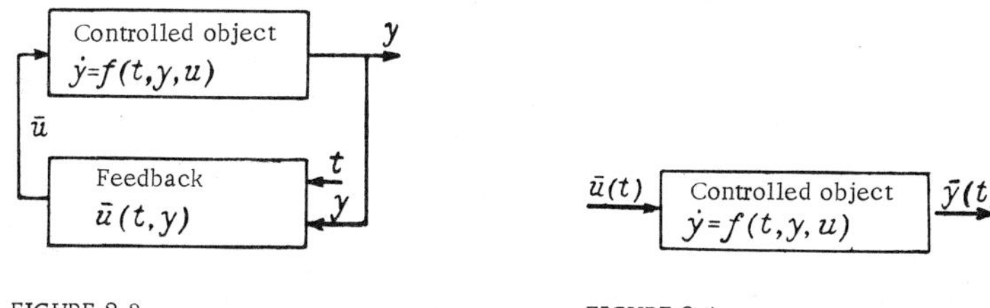

FIGURE 2.3   FIGURE 2.4

The method described above is in fact the Jacobi-Hamilton method whereby variational problems are reduced to partial differential equations. Equation (2.66), apart from the arbitrary function $C(t)$, coincides with Bellman's equation /1/.

Bellman's equation is thus obtained as a particular case from Theorem 1.1 with a function $\varphi(t, y)$ of special form. At the same time, this equation constitutes a sufficient condition of an absolute minimum.

2.2.1. Optimal control synthesis for systems linear in the phase coordinates

Let the right-hand sides of equations (1.3) and the integrand in (1.1) have the form

$$\left.\begin{aligned}
f^k &= \sum_{i=1}^{n} a_i^k(t)\, y^i + h^k(t, u)\ (i, k = 1, 2, ..., n); \\
f^0 &= \sum_{i=1}^{n} a_i^0(t)\, y^i + h^0(t, u),
\end{aligned}\right\} \qquad (2.70)$$

or in vector notation

$$f = A(t)y + h(t, u); \quad f^0 = a^0(t)y + h^0(t, u), \qquad (2.71)$$

where $A(t) = \|a_i^k\|$ is the matrix of the coefficients ($i, k = 1, 2, \ldots, n$). The set V_u is assumed to depend on t alone.

Our object is again to minimize the functional (1.1). The conditions on $V_y(t)$ are the same as in the previous section. Let

$$F(y) \equiv \lambda y, \qquad (2.72)$$

where $\lambda = (\lambda_1, \lambda_2, \ldots, \lambda_n)$ is a given n-dimensional vector; $\lambda y = \sum_{i=1}^{n} \lambda_i y^i$ is the scalar product of the vectors λ and y.

Inserting (2.71) in (1.9), we obtain

$$R(t, y, u) = (A^*\varphi_y - a^0)y + \varphi_t(t, y) + \varphi_y h(t, u) - h^0(t, u),$$

where A^* is the transpose of matrix A, i.e., $A^* = \|a_k^i\|$.

Now from (2.66)

$$\left.\begin{array}{l} P(t, y) = (A^*\varphi_y - a^0)y + \varphi_t + \mathscr{H}(t, y); \\[2mm] \mathscr{H}(t, y) = \sup_{u \in V_u(t)} [\varphi_y h(t, u) - h^0(t, u)]. \end{array}\right\} \qquad (2.73)$$

In order to satisfy equation (2.66), we have to choose $\varphi(t, y)$ so that the function P is independent of y.

We seek φ in the form

$$\varphi = \psi(t)y. \qquad (2.74)$$

Inserting (2.74) in (2.73), we find

$$P = (\dot{\psi} + A^*\psi - a^0)y + \overline{\mathscr{H}}(t); \qquad (2.75)$$

$$\overline{\mathscr{H}}(t) = \sup_{u \in V_u(t)} [\psi(t)h(t, u) - h^0(t, u)]. \qquad (2.75a)$$

The vector function $\psi(t)$ is defined by the equations

$$\left.\begin{array}{l} \dot{\psi} + A^*(t)\psi = a^0(t); \\[2mm] \psi(t_1) = -\lambda. \end{array}\right\} \qquad (2.76)$$

The function $P(t, y) = \mathscr{H}(t)$ is then independent of y and $\Phi(y) \equiv \lambda y + \psi(t_1)y = 0$, i.e., conditions (2.66) and (2.69) imposed on $\varphi(t, y)$ are satisfied. The search for the function $\varphi(t, y)$ is thus reduced to the solution of a Cauchy problem for a system of linear differential equations (2.76) with initial conditions (2.69). If A and a^0 are independent of t, equations (2.76) are integrable in a closed form. Given the vector $\psi(t)$, we find the optimal control function $\bar{u}(t)$ from equation (2.76), which in this case has the form

$$\psi(t)h(t, \bar{u}) - h^0(t, \bar{u}) = \mathscr{H}(t). \qquad (2.75a)$$

Let

$$h^k = b^k u \quad (k = 0, 1, \ldots, n),$$

where u is a scalar (i. e., $r=1$) constrained by the inequality $|u| \leqslant 1$, $b^k(t)$ are continuous functions on $[t_0, t_1]$. We have

$$\mathscr{H}(t) = \sup_{|u| \leqslant 1} [\psi(t)b - b^0]u = |\psi(t)b - b^0|.$$

Here $b = (b^1, b^2, \ldots, b^n)$. We have

$\bar{u}(t) = 1$ for all t when $\psi(t)b - b^0 > 0$;
$\bar{u}(t) = -1$ for all t when $\psi(t)b - b^0 < 0$.

The equation

$$\psi(t)b - b^0 = 0$$

defines the set of "switching points."

The problem of the optimal approach to the hyperplane $t=t_1$ from any position y_0 is thus completely solved for the particular case when f^i $(i=0, 1, \ldots, n)$ are linear functions of the phase coordinates.

2.2.2. Algorithmic features of the method

The Hamilton-Jacobi method reduces the variational problem to the solution of a Cauchy problem for a second-order partial differential equation for the function $\varphi(t, y)$. This problem has a number of characteristic features. Equation (2.66) contains only the partial derivatives φ_y and φ_t, and does not contain the function φ itself. All the partial derivatives, except φ_t, enter the equation nonlinearly, because of the nonlinear operation of the supremum over u. Equation (2.66) is always solvable for the partial derivative φ_t; the boundary conditions are defined on the hyperplane $t=t_1$ in the (t, y) space. Because of the last two factors, the argument t occupies a preferential position among the $n+1$ arguments of the function φ and makes it possible to construct the solution of the problem in the direction of decreasing t, starting with t_1.

The solution algorithm, unlike that of the Lagrange's method, is thus independent of the boundary-value problem. This is one of the principal advantages of the method. Another advantage is that the solution of the partial differential equation (2.66) completely exhausts the solution of the variational problem: any pair $(\bar{y}(t), \bar{u}(t))$ constructed by this method is the absolute minimizing solution, whereas in Lagrange's method the solution of the boundary-value problem for the corresponding system of differential equations only constitutes the first step toward the solution of the problem, as it remains to verify whether or not the extremal obtained in this way is indeed the absolute minimizing solution. Finally, an obvious advantage of the method is that it solves the optimal synthesis problem, which constitutes a much more general problem than the simple search for a single pair $(\bar{y}(t), \bar{u}(t))$.

The greater generality of the result, however, is an outcome of the greater complexity of the algorithm: instead of a system of ordinary differential equations, as in Lagrange's method, we are dealing with a partial differential equation (2.66). No regular numerical methods are available for such an equation. Moreover, no existence theorems for the solutions of equations of this type are known either.

It is therefore desirable to develop suitable methods for approximate optimal synthesis. A general approach to this task and one of the particular schemes are described in the next section.

A shortcoming of this method compared to Lagrange's method is that its application is more restricted. It is not by chance that it is described in reference to a particular class of problems, the so-called problems with a free right end point.

However, numerous problems with a set V_y of a different structure may be approximately reduced to this particular problem by an appropriate modification of the functional. For example, problems with a fixed right end point (fixed y_1) for a functional I are readily reduced to problems with a free right end point for the functional

$$I + \lambda_i \left(y_1^i - y_{1f}^i \right)^2,$$

where y_{1f}^i are the fixed values, y_1^i, λ_i are sufficiently large positive numbers.

§ 2.3. THE METHOD OF APPROXIMATE OPTIMAL CONTROL SYNTHESIS

2.3.1. Statement of the problem

In virtue of the equations

$$\dot{y} = \hat{f}(t, y, u) \tag{2.77}$$

the value of the functional

$$I = \int_{t_0}^{t_1} f^0(t, y, u)\,dt + F(y_1) \tag{2.78}$$

is given if the initial point (t_0, y_0) and the control program $u(t)$ on $[t_0, t_1]$ are given. Let

$$d(t_0, y_0) = \inf_{\{u(t)\}} I[t_0, y_0, u(t)], \tag{2.79}$$

where $\{u(t)\}$ is the set of all the admissible control functions on $[t_0, t_1]$.

A function $u(t)$ is considered admissible if there exists a function $y(t)$ such that the pair $(y(t), u(t)) \in D$. The latter relation implies that the solution of the system of equations

$$\dot{y} = f(t, y, u(t)), \quad y(t_0) = y_0 \tag{2.80}$$

for all $t \in [t_0, t_1]$ belongs to $V_y(t)$ and $u(t) \in V_u(t, y(t))$. It is assumed that $\{u(t)\}$ is not empty. In virtue of the definition of a lower bound, there exists a sequence

$$\{u_\alpha(t)\} \subset \{u(t)\},$$

such that

$$I[t_0, y_0, u_\alpha(t)] \to d(t_0, y_0) \text{ for } \alpha \to \infty.$$

For every fixed initial point (t_0, y_0), we have to select such a sequence from a certain given set V_{y0} in the space (t, y). In other words, we have to solve the previously stated problem for all t_0 and $y_0 \in V_{y0}$.

Let $t_{0\,\min}$ be the minimal t_0 corresponding to the set V_{y0}.

The set $V_y(t)$, $t \in [t_0, t_1]$, in each of the resulting problems is defined as follows: each $t \in [t_{0\,\min}, t_1]$ is assigned to a set $V_{y1}(t)$ and it is assumed that

$$V_y(t_0) = y_0 \in V_{y1}(t_0);$$
$$V_y(t) = V_{y1}(t) \text{ for } t \in (t_0, t_1].$$

Our problem is solved once we have constructed a minimizing sequence of synthesizing functions $u_\alpha(t, y)$ $t \in [t_{0\,\min}, t_1]$, $y \in V_{y1}(t)$. A sequence $u_\alpha(t, y)$ of synthesizing functions is said to be a minimizing sequence if the sequence of the control functions

$$\overline{u}_\alpha(t_0, y_0, t) = u_\alpha[t, \overline{y}_\alpha(t)]$$

belongs to the set of admissible control functions and constitutes a minimizing sequence for any t_0, $y_0 \in V_{y0}$. Here $\overline{y}_\alpha(t)$ is the solution of system (2.80) with the initial conditions $y(t_0) = y_0$ for $u = u_\alpha(t, y)$.

The problem is finally stated in the following form: in a region V_{y0} of the $(n+1)$-dimensional space (t, y), construct a minimizing sequence of control fields $\{u_\alpha(t, y)\}$ for the functional (2.78).

Every element of this sequence will be called an approximate optimal synthesis. The degree of approximation will be measured by the number

$$a_\alpha = \sup_{(t_0, y_0) \in V_{y0}} [I(t_0, y_0, \overline{u}_\alpha(t_0, y_0, t)] - d(t_0, y_0).$$

Evidently, $a_\alpha \to 0$ for $\alpha \to \infty$.

The following theorem enables us to estimate the accuracy of the synthesis without solving equation (2.65).

Consider a function $\varphi(t, y)$ which is continuous and differentiable for $t \in [t_{0\,\min}, t_1]$, $y \in V_{y1}(t)$. Given $\widetilde{\varphi}(t, y)$, we can construct the function

$$\widetilde{R}(t,y,u) = \widetilde{\varphi}_y(t,y) f(t,y,u) - f^0(t,y,u) + \widetilde{\varphi}_t. \tag{2.81}$$

Let $\widetilde{u}(t, y)$ be the control function for which $\widetilde{R}$ attains its maximum value on $V_u(t, y)$, i.e.,

$$\widetilde{R}(t,y,\widetilde{u}(t,y)) = \sup_{u \in V_u(t,y)} \widetilde{R}(t,y,u) \equiv \widetilde{P}(t,y); \tag{2.82}$$

$$\widetilde{\Phi}(y)=F(y)+\widetilde{\varphi}(t_1,y); \qquad m=\inf_{y\,\in V_y(t_1)}\Phi(y); \tag{2.83}$$

$\widetilde{y}(t,\,t_0,\,y_0)$ is the solution of the system

$$\dot{y}=f(t,y,\widetilde{u}(t,y)),\quad y(t_0)=y_0$$
$$\text{and}\qquad\qquad\qquad \widetilde{u}(t,t_0,y_0)=\widetilde{u}(t,\widetilde{y}(t,t_0,y_0)). \tag{2.84}$$

Let for all $(t_0,\,y_0)\in V_{y0}$ and $t\in(t_0,\,t_1)$

$$y(t,\,t_0,\,y_0)\in V_y(t).$$

Let further

$$\widetilde{\Delta}=\int_{t_{0\,\min}}^{t_1}\Big|\sup_{y\,\in V_y(t)}\widetilde{P}(t,y)-\inf_{y\,\in V_y(t)}\widetilde{P}(t,y)\Big|\,dt+\sup_{y\,\in V_y(t_1)}\widetilde{\Phi}(y)-\inf_{y\,\in V_y(t_1)}\widetilde{\Phi}(y). \tag{2.85}$$

T h e o r e m .   The functional (2.78) over the field of control functions $\widetilde{u}(t,\,y)$ satisfies the following estimate:

$$|I[t_0,y_0,\widetilde{u}(t)]-d(t_0,y_0)|\leqslant\widetilde{\Delta}(t_0)\leqslant\widetilde{\Delta}. \tag{2.86}$$

Here $\widetilde{u}(t)=\widetilde{u}(t,\,\widetilde{y}(t))$, $\widetilde{y}(t)$ is the solution of (2.80) for $u=\widetilde{u}(t,\,y)$.
P r o o f .   Consider the functional

$$L=\int_{t_0}^{t_1}R[t,y,\varphi(t,y),u]\,dt-\varphi(t_0,y_0)+\varphi(t_1,y_1)+F(y_1), \tag{2.87}$$

defined over the set E of the independent pairs of vector functions $(y(t),\,u(t))$. This functional has the following property: on the set $D\subset E$ of the pairs $(y(t),\,u(t))$ satisfying equations (2.80), $L=I$ for any $\varphi(t,\,y)$.
Let

$$\widetilde{L}[y(t),u(t)]=L[y(t),u(t),\widetilde{\varphi}(t,y)]; \tag{2.88}$$

$$\widetilde{l}=\inf_{(y(t),u(t))\,\in E}\widetilde{L}[y(t),u(t)]. \tag{2.89}$$

We have

$$\widetilde{l}=-\int_{t_0}^{t_1}\widetilde{\mu}(t)\,dt+\varphi(t_1,y_1)+m, \tag{2.90}$$

where

$$\widetilde{\mu}(t)=\sup_{u\,\in V_u(t)}\widetilde{R}[t,y,\,u]=\sup_{y\,\in V_{y_1}}\widetilde{P}[t,\,y]. \tag{2.91}$$

Seeing that

$$\widetilde{R}[t,\,\widetilde{y},\,\widetilde{u}(t)]=\widetilde{P}[t,\,\widetilde{y}],$$

and using (2.85), (2.87), and (2.90), we obtain

$$\tilde{L}\,[\tilde{y}\,(t),\,\tilde{u}\,(t)]-\tilde{l}\leqslant\int_{t_0}^{t_1}\,[\,-\tilde{P}\,(t,\,\tilde{y}\,(t)\,+\tilde{\mu}(t)]\,dt+\tilde{\Phi}\,[\tilde{y}\,(t_1)]-m\leqslant\tilde{\Delta}. \qquad (2.92)$$

Since

$$(\tilde{y}(t),\,\tilde{u}\,(t))\in D,$$

we have

$$\tilde{L}\,[\bar{y}(t),\tilde{u}\,(t)]=I\,[t_0,y_0,\tilde{u}(t)],$$

and thus

$$I\,[t_0,y_0,\tilde{u}\,(t)]-\tilde{l}\leqslant\tilde{\Delta}. \qquad (2.93)$$

Since $D\subset E$, we see that $d(t_0,\,y_0)\geqslant\tilde{l}$ and, using (2.93), we find

$$|\,I\,[t_0,y_0,\tilde{u}\,(t)]-d\,(t_0,y_0)\,|\leqslant\tilde{\Delta}. \qquad (2.94)$$

Q. E. D.

Corollary. Consider a sequence $\varphi_\alpha\,(t,\,y)$ such that

$$\Delta_\alpha=\Delta(\varphi_\alpha\,)\to0 \qquad (2.95)$$

for $\alpha\to\infty$. The sequence of the control fields $\tilde{u}_\alpha\,(t,\,y)$ is then a minimizing sequence.

If a particular construction of the sequence $\varphi_\alpha(t,\,y)$ satisfying (2.95) is given and the paths $y_\alpha\,(t)$ corresponding to $\varphi_\alpha(t,\,y)$ are seen, in virtue of (2.80), (2.82), and (2.83), to satisfy the condition

$$y_\alpha\,(t,\,t_0,\,y_0)\in V_y(t)\ \text{for}\ t\in[t_0,\,t_1],\ (t_0,\,y_0)\in V_y(t_0),$$

the corollary implies that a regular algorithm for the construction of the control synthesis $\bar{u}_\alpha(t,\,y)$ is also given, which for sufficiently large α is as close as we desire to the optimum control synthesis. Moreover, the theorem provides a specific estimate (2.94) of the closeness of the synthesis $\bar{u}_\alpha\,(t,\,y)$ to the optimal synthesis.

In practice, the proposed method of construction of the minimizing sequence $u_\alpha(t,\,y)$ is implemented as follows: the sequence $\varphi_\alpha\,(t,\,y)$ is constructed so that $\Delta_\alpha\to0$ for $\alpha\to\infty$, and condition (2.94) is verified a posteriori, after the construction of the synthesis $u_\alpha\,(t,\,y)$. For this method of solution to be successful, the set $V_{y1}(t)$ should correspond to a sufficiently large region of the space Y. Also note that if the set $V_{y1}(t)$ in the initial statement of the problem coincides with the entire space Y, it should be replaced by a bounded region to permit searching for the supremum in (2.91) in practice. $V_{y1}(t)$ should remain sufficiently large so that condition (2.94) is satisfied.

Remark. The construction of the sequence $\varphi_\alpha\,(t,\,y)$ satisfying (2.95) may be considered as an approximate method of solution of the partial differential equation (2.82), if $\varphi_\alpha(t,\,y)$ is convergent in some norm. In our formulation, however, the question of the convergence of $\varphi_\alpha(t,\,y)$ does not

arise. The sequence $\varphi_\alpha(t, y)$, and hence $\tilde{u}(t, y)$, may have no limit. The
only requirement is that (2.95) be satisfied by the exact solution and that
Δ_α be sufficiently small for some α when the solution is approximate.

2.3.2. Construction of a minimizing sequence

As a particular example of the construction of a minimizing sequence,
consider the following technique. $\varphi_\alpha(t, y)$ is defined as a polynomial with
known coefficients $\psi(t)$:

$$\varphi_\alpha(t,y)=\sum_{i_1=0}^{l_1-1} (y^1)^{i_1} \left(\sum_{i_2=0}^{l_2-1} (y^2)^{i_2} \left(\cdots \sum_{i_n=0}^{l_n-1} \psi_{i_1,i_2,\ldots,i_n}(t)(y^n)^{i_n} \right) \cdots \right). \tag{2.96}$$

Expression (2.96) for φ_α contains $\alpha = l_1 \cdot l_2 \ldots l_n$ arbitrary continuous
functions $\psi_{i_1 i_2 \ldots i_n}(t)$ which are selected so as to minimize Δ_α in some way.
To this end, we define N supporting curves $y=y_\beta(t)$ which are arranged
in $V_y(t)$ for each t in the following manner. The range of each variable
y^l corresponding to $V_y(t)$ is partitioned into segments by laying off l_1
points on the y^1 axis, l_2 points on the y^2 axis, etc. Hyperplanes
perpendicular to the corresponding axes are passed through these planes.
The intersections of all these hyperplanes give $N=l_1 \cdot l_2 \ldots l_n$ tabular
points. Repeating the same construction for every time t and ensuring
continuous time-variation of the coordinates of the tabular points, we
obtain a family of supporting curves $y=y_\beta(t)$. In practice, the construction
of supporting curves should make use of families of curves which have a
convenient analytical description, such as the family of straight lines
$y(t)=A_1 t+A_2$, the family of parabolas $y=A_1 t^2+A_2 t+A_0$, the family of
polygonal lines, etc. $(4.117 - 4.119)$

Inserting the expression for $\varphi_\alpha(t, y)$ in (2.82) and demanding that (2.82)
and (2.83) are satisfied on the supporting curves only, we obtain the following
system of differential equations and boundary conditions for the functions
$\psi(t)$:

see (4.126)

(4.132)

$$\sum_{i_1=0}^{l_1-1} (y_\beta^1)^{i_1} \left(\sum_{i_2=0}^{l_2-1} (y_\beta^2)^{i_2} \left(\cdots \sum_{i_n=0}^{l_n-1} \dot{\psi}_{i_1,i_2,\ldots,i_n}(t)(y_\beta^n)^{i_n} \right) \right) =$$

$$=K(t) - \mathscr{H}(t, y_\beta, \varphi_y(t, y_\beta)); \tag{2.97}$$

$$\Phi(y_\beta(t_1)) \equiv F[y_\beta(t_1)] + \varphi(t_1, y_\beta(t_1)) = K_1, \tag{2.98}$$

$$\beta = 0, \; l_1, \; l_2, \ldots, l_n = N.$$

Here $\varphi(t, y)$ is defined by (2.96), and $\varphi_{y^l}(t, y)$ is the partial derivative
of (2.96) with respect to y^l.

From the point of view of numerical work, the arbitrary function of
time $K(t)$ and the constant K_1 are conveniently set equal to zero, $K(t)=0$,
$K_1 = 0$. The number of equations (2.97) then coincides with the number of
reference curves. System (2.97) comprises N first-order linear ordinary
differential equations in the unknown functions $\psi(t)$. The functions $y_\beta^l(t)$

entering the coefficients are known. We recall that the subscript β identifies a given vector function $y_\beta(t)$, and the superscript i specifies the particular component of this vector function. The initial conditions for (2.97) are defined by a system of N linear algebraic equations in N unknown functions $\psi(t_1)$. The problem of solving (2.97) for the derivatives is equivalent to the problem of interpolation of the function $-\mathcal{H}(t, y, \varphi_y)$, defined by its values on the supporting curves using the polynomial (2.96). Similarly, the problem of solving (2.98) is equivalent to the interpolation of the function $[-F(y_1, t_1)]$ defined by its values at the points $y_\beta^i(t)$ using the polynomial (2.96).

Our algorithm is built to cope with these problems, since, as can be shown, they are reduced to the solution of a series of one-dimensional interpolation problems using polynomials of degree $l_1 \cdot l_2, \ldots, l_n$ or, in other words, to the solution of a system of linear algebraic equations of order $l_1, l_2, \ldots, l_n$ with non-degenerate matrices (Vandermonde matrices).

The solution of (2.97) with the appropriate boundary conditions yields the functions $\psi_\beta(t)$, which define $\tilde{\varphi}(t, y)$ and hence the approximate control synthesis $\tilde{u}(t, y)$. The synthesis $\tilde{u}(t, y)$ obtained in this way evidently satisfies the estimate (2.94).

In particular problems, to evaluate the estimate (2.95), it is often convenient to supplement the integration of (2.97) for (t_0, t_1) by a simultaneous solution of the equation

$$\dot{\xi} = \sup_{y \in V_y(t)} \widetilde{P}(t, y) - \inf_{y \in V_y(t)} \widetilde{P}(t, y) \tag{2.99}$$

with the following initial condition for $t = t_1$:

$$\xi(t_1) = \sup_{y \in V_y(t_1)} \widetilde{\Phi} - \inf_{y \in V_y(t_1)} \widetilde{\Phi}. \tag{2.100}$$

Evidently,

$$\Delta = \xi(t_0).$$

In the majority of publications on many-dimensional interpolation, the interpolation polynomial is given explicitly or implicitly in the form (2.96), and the reference points are arranged as described above. In /10/, the interpolation problem is solved in general form and general expressions for the polynomial coefficients are given. These expressions, however, are very unwieldy and it is not clear to what extent they will be useful in practical numerical work on computers and how favorably they will compare with information given in the original implicit form by equations (2.97). In numerical solution of (2.97), the derivatives should be determined at every step from explicit and implicit expressions. The number of steps should be sufficiently large. The example that follows shows that a slight modification of the original equations may enable us to choose the reference curves so that the elements of the coefficient matrix of (2.97) are either constants or simple functions of time; the elements of the corresponding inverse matrix can be determined beforehand, in the form of constants or simple functions of time. In this case, the determination of the derivatives at each step involves computations using fairly simple formulae. Since the matrix inversion is performed only once and thus hardly affects the total computer time, the exact form in which the original information about the function $\psi(t)$ is specified is of no particular consequence.

Let us summarize. We described an algorithm for the approximate construction of an optimal control field $\widetilde{u}_\alpha\,(t,\,y)$. An estimate of the closeness of the approximate synthesis $\widetilde{u}_\alpha\,(t,\,y)$ to the optimal synthesis is given by expression (2.94).

The algorithm comprises the following stages.

1. α reference curves $y_\beta\,(t)$ are defined, i. e., α points in a region $V_y(t_1)$ of the space Y for every fixed $t \in [t_0,\,t_1]$.

2. The Cauchy problem is solved for system (2.97) of ordinary differential equations with initial conditions (2.98). System (2.97) should be solved numerically in the direction from t_1 to $t_{0\,\mathrm{min}}$. While constructing the right-hand sides of (2.97), we determine the synthesizing function $\widetilde{u}_\alpha\,(t,\,y)$ from (2.82).

3. The number Δ_α is computed from (2.95) and inequality (2.86) is then applied to determine the closeness of the synthesis $\widetilde{u}_\alpha\,(t,\,y)$ to the optimal synthesis. If the result is insufficiently close, a larger α is selected and the procedure is reiterated.

E x a m p l e 1. Construct the optimal synthesis for the system $\dot{y}=u$ in the region $t<t_1$ of the $t,\,y$ plane from the condition of minimum of the functional

$$I=\int\limits_{t_0}^{t_1} u^2dt+\lambda y^2(t_1),\quad \lambda>0. \tag{2.101}$$

Because of the increment $\lambda y_1{}^2$, the functional falls in a sufficiently small neighborhood of $y(t_1)=0$ for sufficiently large λ. We have

$$R(t,y,u)=\varphi_y u-u^2+\varphi_t;\quad u(t,y)={}^1\!/_2\varphi_y(t,y)$$
$$P(t,y)=\sup_u R(t,y,u)={}^1\!/_4\,\varphi_y^2+\varphi_t;\quad \mathcal{H}(t,y,\varphi_y)={}^1\!/_4\varphi_y^2.$$

The function $F(y)=\lambda y^2$ is a quadratic polynomial, and the least number α for which (2.98) may be satisfied is therefore $\alpha=2$. Let us now construct an approximate synthesis $u_2(t,\,y)$. We have

$$\varphi_2(t,y)=\psi_1(t)\,y+\psi_2(t)\,y^2;$$
$$\frac{\partial}{\partial y}\,\varphi_2=\psi_1(t)+2\psi_2(t)\,y.$$

The curves $y=y_\beta\,(t)$ $(\beta=1,\,2)$ are chosen as the straight lines

$$y_1\equiv 0;\; y_{2,\,3}\equiv \pm 1.$$

1. System (2.97) has the form

$$\left.\begin{aligned}
\dot{\psi}_1+\dot{\psi}_2+\frac{1}{4}\,[\psi_1+2\psi_2]^2-\frac{1}{4}\,\psi_1^2=0;\\[2mm]
-\dot{\psi}_1+\dot{\psi}_2+\frac{1}{4}\,[\psi_1-2\psi_2]^2-\frac{1}{4}\,\psi_1^2=0
\end{aligned}\right\} \tag{2.102}$$

or

$$\dot{\psi}_2 + \psi_2^2 = 0; \quad \dot{\psi}_2 + \psi_2^2 + \dot{\psi}_1 + \psi_1 = 0.$$

The boundary conditions (2.98) have the form

$$\psi_1(t_1) = 0; \quad \psi_2(t_1) = -\lambda. \tag{2.103}$$

The solution of system (2.102), (2.103) is

$$\psi_1(t) \equiv 0; \quad \psi_2(t) = -\frac{1}{t_1 + 1/\lambda - t}.$$

Further,

$$\frac{\partial}{\partial y}\, \varphi_2(t,y) = -\frac{2y}{t_1 + 1/\lambda - t} \; ;$$

$$u_2(t,y) = -\frac{y}{t_1 + 1/\lambda - t} \; . \tag{2.104}$$

Check the closeness of $\ddot{u}_2(t, y)$ to the optimal synthesis:

$$P_2(t,y) = \frac{1}{4}\,(2\psi_2 y)^2 + \dot{\psi}_2 y^2 = [\psi_2^2 + \dot{\psi}_2]\, y^2 = 0, \tag{2.105}$$

i. e., the second approximation coincides with the exact solution.

Example 2. Construct a synthesis for the system

$$\dot{y} = u \quad |u| \leqslant 1 \tag{2.106}$$

in the region $0 \leqslant t \leqslant 0.5$, $|y| < 1$ of the (t, y) plane which will be optimal in terms of ensuring minimum deviation from zero of the coordinate y in the measure

$$I = \int\limits_{t_o}^{t_1} y^2 dt.$$

We have

$$\left.\begin{aligned}
R(t,y,u) &= \varphi_y u - y^2 + \varphi_t; \\
u(t,y) &= \operatorname{sign} \varphi_y;
\end{aligned}\right\} \tag{2.107}$$

$$\left.\begin{aligned}
P(t,y) &= |\varphi_y| - y^2 + \varphi_t; \\
\mathscr{H}(t,y,\varphi_y) &= |\varphi_y| - y^2.
\end{aligned}\right\} \tag{2.108}$$

First approximation:

$$\varphi_1(t, y) = \psi(t) y.$$

Let $y_\beta(t) \equiv \pm 1$, $\beta = 1, 2$.

System (2.97) is written in the form

$$2\dot{\psi} \equiv 0,$$

and the initial condition (2.98) takes the form $\psi(t_1)=0$.   We have

$$\psi(t)\equiv 0; \quad P_1(t,y)=-y^2;$$

$$\Delta_1=\int_0^{0.5}|\sup P_1(y)-\inf P(y)|\,dt=\int_0^{0.5}dt=0.5. \qquad (2.109)$$

Any control function $\bar{u}(t,y)$ is possible by (2.108).   The first approximation in this case thus does not produce a synthesis.   Estimate (2.109) in this case indicates that for any admissible control function, the deviation of the functional from its minimum value does not exceed 0.5.
Second approximation:

$$\varphi_2(t,y)=\psi_1(t)\,y+\psi_2(t)\,y^2;$$
$$\varphi_{2y}=\psi_1+2\psi_2 y.$$

Let

$$y_1(t)\equiv 0; \quad y_{2,3}\equiv \pm 1.$$

System (2.97) takes the form

$$\left.\begin{array}{l}\dot{\psi}_1+\dot{\psi}_2+|\psi_1+2\psi_2|-|\psi_1|-1=0;\\[4pt] -\dot{\psi}_1+\dot{\psi}_2+|\psi_1-2\psi_2|-|\psi_1|-1=0.\end{array}\right\} \qquad (2.110)$$

Initial conditions (2.98) take the form

$$\psi_1(t_1)=0; \quad \psi_2(t_1)=0.$$

This system has the solution

$$\psi_1(t)\equiv 0; \quad \psi_2(t)=1/2\,[e^{2(t-t_1)}-1].$$

Furthermore

$$u_2(t,y)=\operatorname{sign}2\psi_2(t)\,y=-\operatorname{sign}y.$$

It is readily seen that the synthesis $u_2(t,\ y)$ coincides with the optimal synthesis.   Let us check estimate (2.94).   Using the above relations, we find

$$P_2(t,y)=(1-e^{2(t-t_1)})(|y|-y^2);$$

$$\Delta_2=\int_0^{0.5}|\sup_{y\in[-1,+1]}P_2(t,y)-\inf_{y\in[-1,+1]}P_2(t,y)|\,dt=$$

$$=\int_0^{0.5}\frac{1}{4}(1-e^{2(t-t_1)})\,dt=0.045;$$

$$\frac{\Delta_2}{\Delta_1}=\frac{0.045}{0.5}=9\%.$$

Despite the fact that $u_2(t, y)$ is a strictly optimal synthesis, the estimate is not zero, unlike that in Example 1: the a priori estimate in this example is thus too high.

Bibliography for Chapter II

1. Bellman, R. Dynamic Programming. — Princeton Univ. Press. 1957.
2. Boltyanskii, V.G. Matematicheskie metody optimal'nogo upravleniya (Mathematical Methods of Optimal Control). — Nauka. 1966.
3. Vinokurov, V.A. Obobshchennyi metod N'yutona dlya resheniya kraevykh zadach (Generalized Newton's Method for the Solution of Boundary-Value Problems). — Kosmicheskie Issledovaniya, No. 4. 1965.
4. Gantmakher, F.R. Teoriya matrits (Matrix Theory). — Nauka. 1967.
5. Demidovich, B.P. and I.A.Maron. Osnovy vychislitel'noi matematiki (Elements of Numerical Methods). — Nauka. 1965.
6. Krasovskii, N.N. K teorii optimal'nogo upravleniya (Optimal Control Theory). — Avtomatika i Telemekhanika, **18**, No. 11. 1957.
7. Letov, A.N. Analiticheskoe konstruirovanie regulyatorov (Analytical Design of Regulators). — Avtomatika i Telemekhanika, **21**, No. 4—6. 1960; **22**, No. 4. 1961; **23**, No. 11. 1962.
8. Moiseev, N.N. Metody dinamicheskogo programmirovaniya v teorii optimal'nykh upravlenii (Dynamic Programming Methods in Optimal Control Theory). — Zhurnal Vychislitel'noi Matematiki i Matematicheskoi Fiziki, **4**, No. 3. 1964; **5**, No. 1. 1965.
9. Eneev, G.M. O primenenii gradientnogo metoda v zadachakh teorii optimal'nogo upravleniya (Application of the Gradient Method to Optimal Control Theory). — Kosmicheskie Issledovaniya, No. 5. 1963.
10. Stojakovic. Solution du problème d'inversion d'une classe importante de matrices. — C.r.Aca.Sci., 246:1133. 1958.

Chapter III

DEGENERATE PROBLEMS. SLIDING CONTROL

In Chapter II we assumed that an absolute minimizing solution $(\bar{y}(t),\ \bar{u}(t))$ of the functional (1.1) existed in the class D of the admissible pairs of vector-functions $y(t),\ u(t)$. This assumption is not universally true, although it is valid for numerous important problems. The class D may contain no absolute minimizing solutions. There always exists a sequence $\{y_S(t),\ u_S(t)\} \subset D$, however, such that

$$I(y_S, u_S) \xrightarrow[S\to\infty]{} d,\ d = \inf_{(y(t),u(t))\,\in D} I[y(t), u(t)]$$

(a minimizing sequence).

If a minimizing solution $(\bar{y}(t), \bar{u}(t)) \in D$ does not exist, the functional (1.1) is minimized by finding a minimizing sequence. Once a minimizing sequence has been found, we can approximate as close as desired to the optimal control, always remaining in the class of admissible functions.

A construction of a minimizing sequence for the classical object of variational calculus — the functional $\int f^0(t,\ y,\ y')dt$ — is described in /4 — 6/ (see also Supplement). The main feature of this construction is that the sequence of paths $y_S(t)$ in the phase space goes to some function $\bar{y}(t)$ (the zero closeness function), whereas the sequence of control functions $u_S(t)$ has no limit, going to infinitely frequent switching between several (two for a plane problem) fixed control functions $u_\beta(t)$. These terminal functions $\bar{y}(t)$ and $\bar{u}_\beta(t)$, $\beta = 1, 2$, fully describe a minimizing sequence. Sequences of this kind define what is known in automation theory as s l i d i n g   c o n t r o l. Technical examples of a minimizing sequence are provided by the "intermittent thrust control" of an aircraft ensuring maximum range with the engine being switched on and off with the highest possible frequency (see Supplement) of the optimal pulsating punching schedule with the press operating at the highest possible frequency /7/.

In this chapter we will describe some methods of solution of variational problems for the case when no minimizing solution exists in the class of admissible paths, and investigate the properties of minimizing sequences. In terms of the corresponding algorithm, these problems are part of a wider class of so-called degenerate variational problems, whose solution involves a number of specific difficulties. In particular, the methods of the previous chapter are ineffective for the solution of degenerate problems even if the minimizing solution is contained among the admissible pairs $y(t),\ u(t)$. The theory developed in this chapter is in fact the theory of degenerate variational problems.

Let us first consider a particular case of the variational problem advanced in § 1.1, which illustrates the specific difficulties encountered in the solution of degenerate problems.

§ 3.1. A PARTICULAR PROBLEM

Let $f^i, i=0,1,\ldots, n$ be functions of the one variable u, let $V_y(t)$ coincide with the entire space Y for any $t \in (t_0, t_1)$, and let V_u be a closed region independent of t and y. The end points are fixed, $t_0=0;\ t_1=t_{1f};\ y(0)=0;$ $y(t_1)=y_{1f}$.

To solve the problem, we make use of Theorem 1.1. We have

$$R(t,y,u)=\varphi_{y^i}(t,y)f^i(u)-f^0(u)+\varphi_t(t,y). \tag{3.1}$$

If we take $\varphi(t, y)=\psi y$, where $\psi=$ const, the function

$$R=R(u)=\psi_i f^i(u)-f^0(u) \tag{3.2}$$

is independent of y and the sufficient condition of an absolute minimum of the functional on the pair $(\bar{y}(t),\ \bar{u}(t)) \in D$ takes the form

$$R(\bar{u})= \sup_{u \in V_u} R(u). \tag{3.3}$$

Let $R(u)$ have a single supremum point $\bar{u}$ for any ψ. This point is clearly independent of t. Then, by Theorem 1.1, the pair $\bar{y}(t)=f(\bar{u})t,\ \bar{u}(\psi)$ is an absolute minimizing solution of the functional. The vector $\psi=$ const is defined by the condition that the straight line $(t, \bar{y}(t))$ passes through given points $(0, 0)$ and (t_{1f}, y_{1f}) of the (t, y) space. For simplicity, a three-dimensional space is assumed, i. e., $n=2$. If the position of the terminal point (t_{1f}, y_{1f}) is altered, the vector ψ, the optimal control function $\bar{u}$, and the direction of $\bar{y}$ of the straight line $\bar{y}(t)$ all change.

Now suppose that there exists a vector $\psi=(\psi_1, \psi_2)$ such that $R(u)$ has at least $n+1=3$ suprema u_1, u_2, u_3 where

$$\begin{vmatrix} 1 & f^1(u_1) & f^2(u_1) \\ 1 & f^1(u_2) & f^2(u_2) \\ 1 & f^1(u_3) & f^2(u_3) \end{vmatrix} \neq 0. \tag{3.4}$$

$$\sup R(u_1) = \sup R(u_2) = \sup R(u_3)$$

Inserting the control functions u_1, u_2, u_3 in the equation $\dot{y}=f(u)$, we obtain three linearly independent directions in the three-dimensional space (t, y^1, y^2), which may be defined by the vectors

$$a_\beta=[1,\ f^1(u_\beta),\ f^2(u_\beta)],\ \beta=1, 2, 3. \tag{3.5}$$

Any vector in the (t, y) space, including the vector $a(t_{1f},\ y_{1f}^1,\ y_{1f}^2)$, may be represented as a linear combination of the vectors a_β:

$$a=\gamma^\beta a_\beta,\ \beta=1, 2, 3,$$

where $\gamma^1, \gamma^2, \gamma^3$ are some numbers.

In other words, the origin may be joined to any point (t_1, y_1^1, y_1^2) by a polygonal line consisting of straight segments parallel to the vectors a_β. This polygonal line belongs to the class of admissible paths if and only if the coefficients γ^1, γ^2, γ^3 (the projections of the vector a onto a_β) are non-negative:

$$\gamma^\beta \geqslant 0, \ \beta = 1, 2, 3. \tag{3.5a}$$

Condition (3.5) has an obvious geometrical interpretation: the terminal point $(t_{1f}, y_{1f}^1, y_{1f}^2)$ can be joined to the initial point $(0, 0, 0)$ by a polygonal line consisting of segments parallel to a_1, a_2, a_3, which belongs to the class of admissible paths only if this point lies inside the trihedral solid angle (cone) ω with its apex at the origin, spanned by the vectors a_1, a_2, a_3 (Figure 3.1). The situation is entirely similar in the n-dimensional problem.

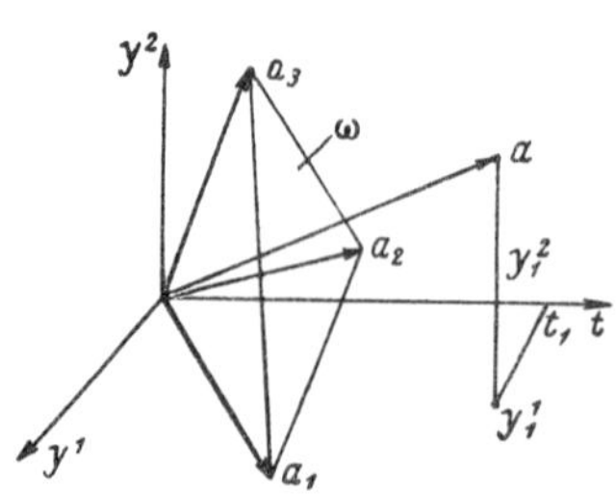

FIGURE 3.1

Each of these polygonal lines is an absolute minimizing solution. It corresponds to a piecewise-constant control function $\bar{u}(t)$ taking on the values u_1, u_2, u_3. Indeed, any such pair $\bar{y}(t)$, $\bar{u}(t)$ belongs to the class D of admissible solutions and satisfies all the conditions of Theorem 1.1. If $(t_1, y_1) \in \omega$, there are obviously infinitely many such minimizing solutions. We can thus always select a minimizing solution consisting of the least number of straight segments (and the least number of controls). If the point (t_{1f}, y_{1f}) lies inside the angle ω, this number is $n+1=3$; if it lies on one of the faces of the trihedral angle, $n=2$, and finally if it lies on one of the edges, $n=1$.

Thus, the minimizing solutions of the functional are of fundamentally different form, dependeing on the properties of the function $R(u)$. Indeed, if $R(u)$ has a single supremum point $\bar{u}$ over $u \in V_u$, the minimizing solution $(\bar{y}(t), \bar{u}(t)) \in D$ is unique. To find this minimizing solution, we have to solve a very simple boundary-value problem: find a vector $\psi = (\psi_1, \psi_2)$ such that the straight line $y = f[\bar{u}(\psi)]t$ passes through the point $(t_{1f}, y_{1f}^1, y_{1f}^2)$. If, however, a vector ψ exists such that $R(u)$ has three supremum points u_β, satisfying the condition $(t_{1f}, y_{1f}) \in \omega$, the functional has infinitely many minimizing solutions, which are polygonal lines. To find the constants ψ_1, ψ_2 and the control functions u_β, we do not have to solve any boundary value problems: they are independent of the terminal point $(t_{1f}, y_{1f}) \in \omega$ and are determined from the finite relation (3.3).

Condition (3.3) in our case contains five equations: three necessary conditions of a local maximum, e.g., $R_u(u_\beta) = 0$, $\beta = 1, 2, 3$, if u_β lie inside V_u, and two necessary conditions of an absolute maximum at each of the points u_β:

$$R(u_1) = R(u_\beta), \ \beta = 2, 3. \tag{3.6}$$

The other conditions included in (3.3) are inequalities. The number k of finite equations (3.6) defining the vector ψ will be called the d e g e n e r a c y of the variational problem. The degeneracy k can be defined as the number of linearly independent vectors a_β (the dimension of the solid angle ω) minus one. The above cases correspond (in the same order) to $k=0$ and $k=n=2$.

An intermediate case, $k=1$, is also possible. Here, for some $\psi=(\psi_1,\ \psi_2)$ $R(u)$ has only two suprema, u_1 and u_2, such that the vector $a=(t_{1f},\ y_{1f}^1,\ y_{1f}^2)$ may be written in the form $a=\gamma^\beta a_\beta$, where $a_\beta=[1,\ f^1(u_\beta),\ f^2(u_\beta)]$, $\gamma^\beta\geqslant 0$, $\beta=1,\ 2$. In this case, ω is a plane angle spanned by a_1 and a_2. The functional has infinitely many minimizing solutions, and the corresponding paths in the $(t,\ y)$ space are polygonal lines consisting of straight segments parallel to a_1 and a_2. In contrast to the case $k=2$, the constants ψ_1, ψ_2 and u_β depend on the position of the terminal point (t_{1f}, y_{1f}^1, y_{1f}^2), although they are "less sensitive" to this factor than for $k=0$. For $k=0$, two boundary conditions had to be satisfied,

$$t_1 \cdot f^i\,[\overline{u}\,(\psi_1,\ \psi_2)]=y_{1f}^i,\ \ i=1,2,$$

and for $k=1$ only one condition is to be satisfied: the vectors a_1, a_2 and $a=(t_{1f}, y_{1f}^1, y_{1f}^2)$ should be coplanar.

§ 3.2. SUFFICIENT CONDITIONS OF AN ABSOLUTE MINIMUM AND THE CONSTRUCTION OF A MINIMIZING SEQUENCE

Let us return to the general variational problem assuming a fixed t_1. The sufficient condition of an absolute minimum for this problem is formulated by Theorem 1.1.

Consider a minimizing sequence $\{y_s(t),\ u_s(t)\}\subset D$ and a function $\varphi(t,\ y)$ which satisfy all the conditions of the theorem. Let us analyze the structure of this sequence depending on the form of the function $R(y,\ u,\ t)$. Let $(\overline{y}(t),\ \overline{u}(t))$ be the point of the $(n+r)$-dimensional space $(y,\ u)$ where $R(t,\ y,\ u)$ attains its maximum for a given t.

Let there be one such point on some segment $(\tau_1,\ \tau_2)\subseteq(t_0,\ t_1)$. Let further $\overline{y}(t)$ be a bounded, continuous, and piecewise-differentiable function and $\overline{u}(t)$ a bounded and piecewise-continuous function. Then, by (1.11), everywhere on this segment

$$y_s(t)\to\overline{y}(t),\ \text{and}\ u_s(t)\to\overline{u}(t). \tag{3.7}$$

Since $\{y_s(t),\ u_s(t)\}\subset D$, a necessary condition for the convergence (3.7) is that $\dot{\overline{y}}=f(t,\ \overline{y},\ \overline{u})$ almost everywhere on $(\tau_1,\ \tau_2)$. Thus, for those points where the function $R(t,\ y,\ u)$ has a single supremum point $(\overline{y},\ \overline{u})$, the minimizing sequence converges to $(\overline{y}(t),\ \overline{u}(t))$, and this pair of vector functions should satisfy equations (1.3).

Now suppose that on the segment $(\tau_1,\ \tau_2)\subset(t_0,\ t_1)$ there are m different control functions u_β such that $R(t,\ y,\ u)$ has a supremum at every point $(\overline{y},\ u_\beta)$ of the space $(y,\ u)$ ($\overline{y}$ is the same for all the suprema), i.e.,

$$R\,[t,\ \overline{y}(t),\ u_\beta(t)]=\mu(t),\ \beta=1,2,\ldots,m, \tag{3.8}$$

and let there be $k+1 \leqslant m$ control functions u_β, $\beta=1, 2, \ldots, k+1$, such that the vectors $a_\beta = [1, f(t, \bar{y}, u^\beta)]$ of the (t, y) space are linearly independent. Then, by (1.11), $\bar{y}_S(t) \rightarrow \bar{y}(t)$, whereas the control function $u_S(t)$ does not necessarily have a limit. It is sufficient if this control function "oscillates" between $u_\beta(t)$. The latter signifies that for any $\varepsilon > 0$ and a sufficiently large terminal S, one of the inequalities

$$|u_S(t) - u_\beta(t)| < \varepsilon, \quad \beta = 1, 2, \ldots, m, \tag{3.9}$$

should be satisfied everywhere on (τ_1, τ_2), with the exception of a set of points whose measure goes to zero for $S \rightarrow \infty$.

The limit function $\bar{y}(t)$ in this case is not necessarily a solution of (1.3) for any $u(t)$. It is necessary, however, that on (τ_1, τ_2) the vector $a(t) = (1, \dot{\bar{y}})$ of the space (t, y) belongs to the angle $\omega[t, \bar{y}(t)]$, which is a convex envelope of the vectors

$$a_\beta = [1, f^1(t, \bar{y}, u_\beta), f^2(t, \bar{y}, u_\beta), \ldots, f^n(t, \bar{y}, u_\beta)],$$
$$\beta = 1, 2, \ldots, k+1.$$

In other words, it is necessary that there exist some functions $\gamma^\beta(t)$, $\beta = 1, 2, \ldots, k+1$, such that on (τ_1, τ_2) the vector $(1, \dot{\bar{y}})$ is representable in the form

$$a(t) \equiv (1, \dot{\bar{y}}) = \gamma^\beta \cdot a_\beta; \quad \gamma^\beta(t) \geqslant 0, \tag{3.10}$$

$$\dot{\bar{y}} = \gamma^\beta(t) f(t, \bar{y}, u_\beta), \tag{3.11}$$

or

$$\sum_\beta^{k+1} \gamma^\beta = 1; \quad \gamma^\beta \geqslant 0, \quad \beta = 1, 2, \ldots, k+1 \tag{3.12}$$

(summation over β is implied). The functions $\gamma^\beta(t)$ are piecewise-continuous. As a result, $\bar{y}(t)$ is piecewise-differentiable.

The higher the dimension $k+1$ of the angle ω, the weaker the "coupling" imposed by (1.3) on the vector function $\bar{y}(t)$. For $k=n$, the condition $(1, \dot{\bar{y}}) \in \omega$ imposes simple inequality constraints on $\bar{y}(t)$, instead of equations (1.3). For $k=0$, conversely, $\bar{y}(t)$ should be the solution of the equation

$$\dot{y} = f[t, y, u_\beta(t)],$$

where u_β is any of the maxima of the function R for a fixed $y = \bar{y}(t)$.

We will now show that if $(1, \dot{\bar{y}}) \in \omega$ on (τ_1, τ_2), we may indeed construct a sequence $\{y_S(t), u_S(t)\} \subset D$ satisfying (1.3) and (1.11) on (τ_1, τ_2). Moreover, we will actually construct this sequence. We partition the segment (τ_1, τ_2) into S intervals by the points $\tau_1 = t^0 < t^1 < \ldots < t^S < \tau_2$. This partition is expected to satisfy one condition only: for $S \rightarrow \infty$,

$$\Delta_S = \max_{\text{for } S \to \infty} |t^{\gamma+1} - t^\gamma| \rightarrow 0, \quad \gamma = 0, 1, 2, \ldots, S-1.$$

Let

$$y_s(t^\tau) = \bar{y}(t^\tau) + 0(\Delta_S),$$

$$\text{when } \frac{0(\Delta_S)}{\Delta_S} \to 0 \text{ for } \Delta_S \to 0.$$

Every point $(t^\tau, y_s(t^\tau))$ in the (t, y) space is joined with the adjacent point $(t^{\tau+1}, y_s(t^{\tau+1}))$ by a polygonal line consisting of straight segments parallel to the vectors

$$a_\beta(t^\tau) = \{1, f[t^\tau, \bar{y}(t^\tau), u_\beta(t^\tau)]\}. \tag{3.13}$$

$$\beta = 1, 2, \dots, k+1.$$

The polygonal line $y_s(t)$ constructed in this way is a solution, to terms of the order $0(\Delta_S)$, of equation (1.3) with a piecewise-constant control function $u_s(t)$, which is equal to $u_\beta(t^\tau)$ over the part $\gamma^\beta(t^\tau)\Delta_\tau$ of each interval $\Delta_\tau = t^{\tau+1} - t^\tau$. If (1.3) is to be exactly satisfied, the straight segments parallel to the vectors $a^\beta(t^\tau)$ should be replaced by the solutions of the equations

$$\dot{y} = f[t, y, u_\beta(t^\tau)].$$

The sequence $\{y_s(t), u_s(t)\}$ constructed in this way satisfies all the conditions of the theorem on (τ_1, τ_2): it satisfies equations (1.3) and for any $t \in (\tau_1, \tau_2)$,

$$R(t, y_s, u_s) \underset{S \to \infty}{\to} \mu(t),$$

where

$$\mu(t) = R[t, \bar{y}(t), u_\beta(t)]$$

is bounded and continuous almost everywhere on (τ_1, τ_2) in virtue of the properties of the functions R, $\bar{y}(t)$, $u_\beta(t)$. Thus, if this sequence satisfies the conditions of the theorem on $[t_0, t_1]$, i. e., beyond the limits of the interval (τ_1, τ_2), it is a minimizing sequence.

The minimizing sequence constructed by this method has much in common with the sliding control functions of relay systems in automatic control theory, which switch from one state to another. For this reason, it will be referred to as the optimal sliding control.

The limit function $\bar{y}(t)$ of a sequence of optimal paths $y_s(t)$ will be called the zero closeness function of the sliding control. In the limit, the sliding control may be treated as "sliding motion" along the path $\bar{y}(t)$ with infinitely frequent switching between different control functions $u_\beta(t)$.

The control switching may occur in any order, as long as the approximating path $y_s(t)$ remains in a sufficiently small neighborhood of the path $\bar{y}(t)$.

The optimal sliding control is fully defined by the zero closeness function $\bar{y}(t)$ and a set of $k+1$ basis control functions $u_\beta(t)$, $\beta = 1, 2, \dots, k+1$. To find this control, we have to find its defining characteristics.

The least number l of switchings of $u_s(t)$ on every small interval $\Delta_\tau = (t^\tau, t^{\tau+1}) \in (\tau_1, \tau_2)$ will be called the b r a n c h i n g of the optimal control on (τ_1, τ_2). Branching can be defined as the least number of straight segments which make up an elementary polygonal line in the (t, y) space minus one. We have an obvious inequality $l \leqslant k$.

Note that in our construction of the minimizing sequence, we dealt not with the total number m of the suprema $(\bar{y}, u_\beta)$ of $R(t, y, u)$, but with the number $k+1 \leqslant m$ of the basis control functions u_β such that

$$R(t, \bar{y}, u_\beta) = \mu(t), \quad \beta = 1, 2, \ldots, k+1, \tag{3.14}$$

and the vectors $a_\beta [1, f(u_\beta)]$ in the (t, y) space are linearly independent. Condition (3.14) is equivalent on (τ_1, τ_2) to condition (1.11) if the sequence $[y_s(t), u_s(t)]$ is constructed by the above method.

The number k is called the degeneracy of the solution of the variational problem on (τ_1, τ_2). This number, as we have noted before, characterizes to what extent the zero closeness function is "independent" of the constraints (1.3). The control functions $u_\beta(t)$ defining the basis vectors a_β, $\beta = 1, 2, \ldots, k+1$, which span the angle $\omega(t)$ will be called the basis functions of the optimal sliding control.

Let

$$H[t, \bar{y}(t), \psi(t), u] = \psi(t) f[t, \bar{y}(t), u] - f^0(t, \bar{y}, u), \tag{3.15}$$

where

$$\psi(t) = \varphi_y[t, \bar{y}(t)]. \tag{3.16}$$

By (3.14),

$$H[t, \bar{y}(t), \psi(t), u_\beta] = \mu_1(t) \equiv \sup_{u \in V_u(t, \bar{y})} H[t, \bar{y}(t), \psi(t), u], \tag{3.17}$$

$$\beta = 1, 2, \ldots, k+1$$

or

$$\psi f[t, \bar{y}(t), u_\beta] - f^0[t, \bar{y}(t), u_\beta] = \mu_1(t), \tag{3.18}$$

$$\beta = 1, 2, \ldots, k+1.$$

Since the vectors a_β are linearly independent, the matrix

$$\left\| \begin{array}{c} 1 \quad f^1(u_1), \ldots, f^n(u_1) \\ \cdot \cdot \cdot \cdot \cdot \cdot \cdot \cdot \cdot \\ 1 \quad f^1(u_{k+1}), \ldots, f^n(u_{k+1}) \end{array} \right\| \tag{3.19}$$

is of rank $k+1$. This matrix coincides with the matrix of the coefficients before μ_1, ψ in (3.18).

A degeneracy of k thus corresponds to the existence of k independent finite equations (3.18) to be satisfied by the vector $\psi(t)$ and the function $\mu_1(t)$. If on (τ_1, τ_2) the solution has a degeneracy k, then a sliding control with branching $l = k$, a zero closeness function $\bar{y}(t)$, and basis control functions $u_\beta(t)$, $\beta = 1, 2, \ldots, k+1$, is optimal on this interval.

Is this sliding control unique or do there exist other optimal sliding controls with the same zero closeness function but a smaller branching on (τ_1, τ_2)? A sufficient condition for the existence of such controls is the existence of an angle $\omega_1[t,\bar{y}(t)]$ of dimension $k_1+1<k+1$. In particular, for the zero closeness function $\bar{y}(t)$ to be part of the phase path of an absolute minimizing solution, it is sufficient that there exists a control function $v(t) \in V_u$ which, first, satisfies the equation $\dot{\bar{y}}=f(t, \bar{y}, v)$ on (τ_1, τ_2) (the angle ω_1 in this case $k_1=0$ degenerates into the vector $(1, \dot{\bar{y}})$ and, second, satisfies the condition

$$H[t, \bar{y}(t), \psi(t), v]=H[t, \bar{y}(t), \psi(t), u_\beta(t)]=\mu_1(t). \qquad (3.20)$$

For example, in problems which are linear in control functions,

$$f^i = A^i(t,y)+B^i_j(t,y)u^j, \quad i=0,1,2,...n,$$

$$j=1,2,...r,$$

such a solution always exists, and a minimizing solution $(\bar{y}(t), \bar{u}(t)) \in D$ therefore also always exists, although along a so-called singular curve $\bar{y}(t)$ satisfying the equalities

$$\psi_i(t) B^i_j\ [t,\bar{y}(t)] - B^0_j[t,\bar{y}(t)]=0, \qquad (3.21)$$

$$j=1,2,...,q,$$

the degeneracy is $k=q$. Indeed, by (3.21), the function

$$H[t,\bar{y}(t), \psi(t), u]=\psi_i B^i_j u^j - B^0_j u^j +\psi_i A^i((t, \bar{y}(t)) - A^0(t, \bar{y}(t)), \qquad (3.22)$$

$$i=1,2,...,n,$$

$$j=q+1,...r$$

is independent of the control functions u^j, $j=1, 2, ..., q$.

Therefore, any control function $v(t) = \{u^1, ..., u^r\}$, where the first components are arbitrary and the last components maximize (3.22), satisfies (3.20). In particular, a control function with first q components satisfying (1.3) also satisfies (3.20).

Irrespective of the degeneracy k, the minimizing solution $(\bar{y}(t), \bar{u}(t))$ exists in the class of admissible solutions D if the branching l is zero. If on some (τ_1, τ_2), $l>0$, no minimizing solution exists in D, and a minimizing sequence has to be constructed as described above. This solution of the variational problem will be called an optimal control with branching on (τ_1, τ_2), or an optimal sliding control on (τ_1, τ_2), or simply optimal control if there is no danger of confusion.

Theorem 1.1 reduces the problem of minimizing a functional on a set D to the problem of finding a maximum of the function $R(t, y, u)$ for every fixed t. Our unknowns in this problem are the minimizing sequence, defined by the zero closeness function $\bar{y}(t)$ and the basis control functions

u_β, and the function $\varphi(t, y)$. The conditions of the theorem do not prescribe an unambiguous choice of the function $\varphi(t, y)$. Making use of this ambiguity, as in Chapter II, we may select different algorithms based on Theorem 1.1 for the solution of the variational problem.

§ 3.3. GENERALIZATION OF THE LAGRANGE–PONTRYAGIN METHOD

For simplicity, we assume that $\bar{y}(t)$ lies inside V_y for all $t \in (t_0, t_1)$, the vectors $y_0 = y(t_0)$, $y_1 = y(t_1)$ are given, the region V_u is defined by the inequalities

$$|u^j| \leqslant 1, \ j = 1, 2, \ldots, r,$$

and the functions $f^i(t, y, u)$, $i = 0, 1, \ldots, n$, are continuous and twice differentiable.

Consider an optimal control with a zero closeness function $\bar{y}(t)$ which satisfies all the conditions of Theorem 1.1, i.e., equality (1.11) in our case. On $(\tau_1, \tau_2) \subset (t_0, t_1)$, this optimal control has a degeneracy k. We will write out all the equations which are contained in (3.14) and which should be satisfied by the sought functions on (τ_1, τ_2). The function $\varphi(t, y)$ is assumed to be twice differentiable in the process.

1. Condition (3.17) containing k equations (3.18) and $r(k+1)$ conditions for a local maximum of H:
either

$$H_{u^j} = 0, \quad |u_\beta^j| < 1, \tag{3.23}$$

or

$$u_\beta^j = \pm 1, \tag{3.24}$$

where

$$\beta = 1, 2, \ldots, k+1; \ j = 1, 2, \ldots, r.$$

2. The necessary conditions for a maximum of R with respect to y:

$$R_{y^i}(t, \bar{y}, u_\beta) = \varphi_{yy^i} f(t, y, u_\beta) + \varphi_{ty^i} + \varphi_y f_{y^i}(t, y, u_\beta) - f_{y^i}^0(t, y, u_\beta)\big|_{y=\bar{y}(t)} = 0,$$

or

$$R_{y^i}(t, \bar{y}, u_\beta^\beta) = \frac{d}{dt}\left(\varphi_{y^i}^\beta\right) + H_{y^i}(t, \varphi_y, y, u_\beta)\bigg|_{y=\bar{y}(t)} = 0. \tag{3.25}$$

The index β in the first term in (3.25) signifies that the total derivative of $\varphi_{y^i}(t, y)$ is taken in virtue of the equation $\bar{y} = f(t, y, u_\beta)$. The total number of equations in (3.25) is $(k+1)n$.

Using (3.11), we may write

$$\dot{\psi}_i(t) \equiv \frac{d}{dt}\varphi_{y^i}[t, \bar{y}(t)] = \gamma^\beta \frac{d}{dt}[\varphi_y^i]^\beta.$$

For every fixed i, we multiply each of the $k+1$ equalities in (3.25) by γ^β and add them up. Using the last identity, we obtain

$$\dot\psi_i + \gamma^\beta H_{y^i}(t,\psi,\bar y,u_\beta) = 0, \qquad (3.26)$$

$$i=1,2,...,n; \quad \beta=1,2,...,k+1.$$

Let us write the necessary conditions for a maximum of $R(t, y, u)$ on (τ_1, τ_2) supplementing them with conditions (3.11), (3.12) which signify that the sought minimizing sequence belongs to the set of sliding controls on (τ_1, τ_2):

$$\left.\begin{aligned}
&\dot y = \gamma^\beta f(t,y,u_\beta); \\
&\dot\psi + \gamma^\beta H_y(t,y,u_\beta)=0; \\
&H[t,\psi(t),\bar y(t), u_\beta] = \sup_{u\in V_u} H(t,\psi(t), \bar y(t),u); \\
&\sum_{\beta=1}^{k+1} \gamma^\beta = 1; \quad \gamma^\beta(t) \geqslant 0; \quad \beta=1,2,...,k+1.
\end{aligned}\right\} \qquad (3.27)$$

Here H is defined by (3.15) and (3.16).

This system is an analog of the equations of Pontryagin's maximum principle for a minimizing solution of class D. In case of zero degeneracy $(k=0)$, equations (3.27) coincide with (1.3), (2.4), and (2.5) of the maximum principle.

Equations (3.27) contain $2n+r+k+k\cdot r+1$ unknown functions: these include n phase coordinates $y^i(t)$, n functions $\psi_i(t)$, $r(k+1)$ components of basis control functions u_β, and $k+1$ factors $\gamma^\beta(t)$. The number of equations is also $2n+r+k+k\cdot r+1$. Indeed, we have $2n$ equations from (3.11) and (3.26), one equation from (3.12), and $k+r+r\cdot k$ equations from (3.17).

The number of equations in (3.27) is thus equal to the number of unknowns.

Equations (3.27) include k independent finite relations

$$H\left(t,\bar y,\psi,u_\beta(t,\bar y,\psi)\right) - H\left(t,\bar y,\psi,u_1(t,\bar y,\psi)\right) = 0, \qquad (3.28)$$

$$\beta=2,3,...,k+1,$$

which follow from (3.17). These relations indicate that the function $H(t, \bar y, \psi, u)$ for fixed t, y, ψ has equal values at all the suprema u. Hence it follows that the initial conditions (t, y, ψ) for the $2n$ differential equations contained in (3.27) cannot be chosen arbitrarily: they must satisfy (3.28), and so the order of the system (3.27) is at most $2n-k$.

Conditions (3.28) should be satisfied identically along any solution of (3.11), (3.26) on (τ_1, τ_2). Setting the total derivatives of (3.28) with respect to t equal to zero and insering for $\dot\psi$ and $\dot y$ the right-hand sides from (3.11) and (3.26), we obtain the following system of linear equations in γ^β:

$$\gamma^\beta \left[H_y \left|{}^{u_\alpha}_{u_1} f(u_\beta) - H_y f \left|{}^{u_\alpha}_{u_1}\right.\right.\right] = -H_t \left|{}^{u_\alpha}_{u_1},\right. \tag{3.29}$$

$$\alpha = 2,3,...,k+1; \quad \beta = 1,2,...,k+1.$$

Here

$$f\left|{}^{u_\alpha}_{u_1}\right. = f(u_\alpha) - f(u_1), \quad f(u_\alpha) = f(t,y,u_\alpha),$$

etc. The left-hand sides contain scalar products of the n-dimensional vectors f and $H_y = (H_{y^1}, H_{y^2}, ..., H_{y^n})$.

If equations (3.29), (3.12) are compatible for any t, y, ψ, they define the coefficients γ^β. However, these equations may prove to be compatible only for those t, y, ψ which satisfy certain finite relations. This will evidently lower the order of system (3.27) even further.

Let $k=1$. Then (3.28), (3.29), and (3.12) take the form

$$H(t,y,\psi,u_2) - H(t,y,\psi,u_1) = 0; \tag{3.30}$$

$$\gamma^1 \left[(H_y(u_2) - H_y(u_1))f(u_1) - H_y(u_1)(f(u_2) - f(u_1))\right] +$$

$$+ \gamma^2 \left[(H_y(u_2) - H_y(u_1))f(u_2) - H_y(u_2)(f(u_2) - f(u_1))\right] = -H_t(u_2) + H_t(u_1); \tag{3.31}$$

$$\gamma^1 + \gamma^2 = 1. \tag{3.32}$$

Collecting similar terms in (3.31), we obtain

$$(\gamma^1 + \gamma^2)\left[H_y(u_2)f(u_1) - H_y(u_1)f(u_2)\right] = -H_t(u_2) + H_t(u_1),$$

whence, using (3.12), we obtain an additional finite relation supplementing (3.30)

$$\left[H_y(u_2)f(u_1) + H_t(u_2)\right] - \left[H_y(u_1)f(u_2) + H_t(u_1)\right] = 0,$$

so that system (3.27) in this case is at most of order $2n-2$.

The general procedure for the construction of a solution by Lagrange's method is not very clear at this stage. In principle, however, it can be described as follows.

Choose an arbitrary initial vector $\psi^0 = \psi(t_0)$ and investigate the maxima of the function

$$H(t_0,\ y_0,\ \psi^0,\ u).$$

Suppose that $H(t_0, y_0, \psi^0, u)$ has a single absolute maximum u_1. This means that for the particular ψ^0 chosen, the degeneracy is zero at the corresponding point, $k=0$, and equations (3.27) coincide with the ordinary equations of Pontryagin's maximum principle. Using these equations, we construct (numerically, step by step) the functions $\bar{y}(t), \psi(t), \bar{u}_1(t)$. If for some $t=\tau_1$ the function $H[t, \bar{y}, \psi(t), u]$ is found to have several absolute maxima (basis control functions) $u_1(t), ..., u_m(t), m>1$, and system (3.27) is compatible at this point, the construction of the extremal can be continued, in general, from this state in m different directions,

$$(t,\ \dot{y}) = (t,\ f(u_\beta)),$$

taking $k=0$, as before, and a sliding control of degeneracy $k=m-1$ can
be constructed using equations (3.27). Each of these alternatives satisfies
equations (3.27) equally well, but these equations only provide the necessary
conditions for a maximum of $R(t, y, u)$. To choose the optimal alternative,
we must rely on conditions (1.11).

Suppose that the optimal branch on (τ_1, t_1) is a sliding control of
degeneracy $k=m-1$. Equations (3.27) together with the initial conditions
$y=y(\tau_1)$, $\psi=\psi(\tau_1)$ fully define this control. It is constructed as follows.

Given $\bar{y}(\tau_1)$, $\psi(\tau_1)$, we find the absolute maximum u_β of $H(\tau_1, u)$; from
(3.29) and (3.12) we find the factors γ^β. Inserting these $u_\beta(\tau_1)$ and $\gamma^\beta(\tau_1)$
in (3.11) and (3.26), we obtain $\dot{y}(\tau_1), \dot{\psi}(\tau_1)$, and applying any suitable
numerical method of integration of differential equations, we construct
step by step a solution of (3.27) up to $t=t_1$. If $y(t_1)\neq y_{1f}$, a different ψ^0 is
selected and the entire process is repeated; alternatively, for the same ψ^0,
some $\tau_2 \in [\tau_1, t_1]$ is chosen and, if possible, one of the above control
alternatives with degeneracy $k<m-1$ is constructed from this point.

A characteristic feature of system (3.27) with $k>0$ is that its solutions
are not unique. The number of solutions may be infinite, as Example 2.2
in Chapter II shows (a degenerate problem with linear control).

It should be stressed that in the degenerate case, equations (3.27) do not
yield the necessary information even for constructing local minimizing
solutions. With nondegenerate systems, on the other hand, the equations
of the maximum principle are often solvable or have a finite number of
solutions, so that in principle the minimizing solution can be found by
constructing all the possible extremals and choosing the best alternative.

For $k\geqslant 2$, equations (3.29) are often solvable for γ^β, so that (3.27) is
of order $2n-k$, i. e., the set of its solutions contains $2n-k$ constants.
Further analysis of the function R (see § 3.5) shows that the necessary
conditions for a maximum of the second derivative of R are satisfied only
by those solutions of (3.27) (if they exist) for which the coefficients of (3.29)
identically vanish, so that the right-hand sides of this system are all equal.
This generates $\frac{k(k-1)}{2}+k$ additional finite relations. If they turn to be independent, the overall order of system (3.27) does not exceed
$$2n-2k-\frac{k(k-1)}{2}.$$

The system of finite relations may contain less than $2k+\frac{k(k-1)}{2}$ independent relations.

For example, it can be shown that if the degeneracy is $k=n$ and the
integrand is independent of the control function and has a stationary point
$\bar{y}(t)$ for every $t \in (t_0, t_1)$, where $\bar{y}(t)$ is a continuous and differentiable
vector function, system (3.27) is of order zero.

§ 3.4. REMARKS CONCERNING THE HAMILTON – JACOBI – BELLMAN METHOD

The Hamilton–Jacobi–Bellman method was considered in Chapter II for
those cases when the minimum exists in D. Unlike the Lagrange method,

where Pontryagin's maximum principle equations are replaced by (3.27),
the Hamilton—Jacobi—Bellman method is extended to the case of sliding
control without any change.

Indeed, consider the function $P(t, y)$, (2.65), and choose $\varphi(t, y)$ so that
P becomes independent of y, i. e., so that equation (2.66) is satisfied.

As in Chapter II, the function $\varphi(t, y)$ describing the optimal control field
is defined by the partial differential equation (2.66) with the boundary
condition (2.69).

Having found a solution $\tilde{\varphi}(t, y)$ of this equation, we obtain an optimal
control function $\tilde{u}(t, y)$ at every point (t, y), but this control is not
necessarily unique. It is defined by equation (2.65) and in practice is obtained
in the course of construction of $P[t, y, \varphi(t, y)]$.

To obtain an optimal solution corresponding to some fixed initial
condition $(t_0, y_0)_f$, it is sufficient to integrate system (1.3), closed by
equation (2.65), with this initial condition. If the solution of the problem
with this initial condition contains a sliding control section in (τ_1, τ_2), the
required solution is realized automatically. Otherwise we would have ended
up with a non-optimal solution satisfying the conditions of Theorem 1.1,
which is impossible.

In numerical integration, the sliding control section emerges as a
control schedule with very frequent control switching. The switching
frequency increases as the integration interval diminishes, and the sought
minimizing sequence is thus indeed obtained by reducing the integration
interval.

We would like to stress some characteristic features of the partial
differential equation (2.66) in this case.

Let the function $H(t, y, \psi, u)$ have the form

$$H = K(t, y, \psi) + L(t, y, \psi) u,$$

where $u \in [-1, +1]$ is a scalar. The function P thus takes the form

$$P = K(t, y, \varphi_y) + |L(t, y, \varphi_y)| + \varphi_t,$$

i. e., the left-hand side of the partial differential equation (2.66) is non-
differentiable. This situation is characteristic of all degenerate problems.

§ 3.5. THE METHOD OF MULTIPLE MAXIMA

Consider another method of solving degenerate variational problems,
which differs from the Lagrange and the Hamilton—Jacobi methods. When
applicable, it has a number of distinct advantages which will be considered
below. This method uses a particular definition of the function $\varphi(t, y)$.
First let us consider a particular problem of minimizing the functional
(1.1).

3.5.1. The simplest functional

Let

$$I = \int_{t_0}^{t_1} f^0(t,y,u)\,dt;$$

$$\dot{y} = f(t,y,u); \quad \Gamma_1 \leqslant u \leqslant \Gamma_2. \tag{3.33}$$

Here $y(t)$ and $u(t)$ are scalar functions,

$$y_0 = y_{0f}, \quad y_1 = y_{1f}.$$

Let $f_u \geqslant 0$ for $u \in [\Gamma_1, \Gamma_2]$ and let the region $V_y(t)$ of the admissible values of y be restricted to the solutions of the equations $\dot{y} = f(t, y, \Gamma_{1,2})$ which pass through the given initial and terminal points. Additional conditions may also be imposed on $V_y(t)$.

At every point $t, y \in V_y(t)$ we find $\psi = \bar{\psi}(t, y)$ such that $H = \psi f - f^0$ attains a supremum at least for two values of u,

$$H[t,y,\psi(t,y), u_1(t,y)] = H[t,y,\psi,\psi(t,y),u_2(t,y)] = d(t,y); \tag{3.34}$$

$$d(t,y) = \sup_{\Gamma_1 < u < \Gamma_2} H(t,y,\psi(t,y),u), \tag{3.35}$$

and the vectors $[1, f(u_1)]$ and $[1, f(u_2)]$ are linearly independent, i.e., $f(u_1) \neq f(u_2)$.

As we have seen before, relations (3.34) and (3.35) incorporate three equations in three unknowns u_1, u_2, ψ. Let $\Gamma_1 < u_1 < \Gamma_2$ and $u_2 = \Gamma_2$. Then

$$H(t,y,\psi,u)\big|_1^2 \equiv$$

$$\equiv \psi[f(\Gamma_2) - f(u_1)] - [f^0(\Gamma_2) - f^0(u_1)] = 0;$$

$$H_u(u_1) \equiv \psi f_u(u_1) - f_u^0(u_1) = 0. \tag{3.36}$$

Solving the second equation in (3.36) for ψ and inserting the result in the first equation, we obtain an equation for u_1.

The function $\varphi(t, y)$ is now defined by the equality

$$\varphi_y = \psi(t,y);$$

$$\varphi(t,y) = \int \psi(t,y)\,dy. \tag{3.37}$$

The integral in (3.37) is indefinite, with fixed t, since an additive constant in the expression for φ does not affect the solution of the problem. Thus

$$R(t,y,u) = \psi(t,y)\,f(t,y,u) - f^0(t,y,u) + \int \psi_t(t,y)\,dy, \tag{3.38}$$

$$P(t,y) \equiv \sup_u R = d(t,y) + \int \psi_t(t,y)\,dy. \tag{3.39}$$

Maximizing $P(t, y)$ with respect to y for every fixed $t \in (t_0, t_1)$, we find a solution $\bar{y}(t)$ of the equation

$$P(t,\bar{y}) = \sup_{y \in V_y(t)} P(t,y) \equiv \mu(t). \qquad (3.40)$$

The function $\bar{y}(t)$ in general consists of pieces of the boundary of $V_y(t)$ and the solutions of the equation

$$R_y(t,y) = 0 \qquad (3.41)$$

in $V_y(t)$.

Let $\bar{y}(t)$ be continuous and piecewise-differentiable on $[t_0, t]$. Let it further satisfy the inequality

$$\dot{f}(t, \bar{y}, u_1) \leqslant \dot{\bar{y}} \leqslant \dot{f}(t, \bar{y}, u_2), \qquad (3.42)$$

which in our case is equivalent to the condition $(1, \dot{\bar{y}}) \in \omega(t, \bar{y}(t))$. Then $\bar{y}(t)$ is a zero closeness function of an optimal control of degeneracy $k=1$ with the basis control functions u_1, u_2.

Indeed, for every $t \in (t_0, t_1)$, the function $R(t, y, u)$ defined by (3.38) has an absolute maximum at the points $(\bar{y}, u_1)$ and $(\bar{y}, u_2)$ of the plane (y, u). The problem is thus solved.

The above discussion also reveals the characteristic features of this algorithm. In a sense, it is "antipodal" to the Hamilton—Jacobi—Bellman method. Both these methods, in contrast to the Lagrange method, impose certain constraints on $\varphi(t, y)$ which should be satisfied identically in V_y. In Bellman's method the optimal path is selected by choosing a control function $u = \bar{u}$ which ensures $\sup_{V_u} R$, while R is maximized with respect to y by an appropriate choice of $\varphi(t, y)$. In the new method, on the other hand, a minimizing solution is obtained by choosing the point $y = \bar{y}$ in the phase space for every t on which $\sup R$ is attained; the maximum with respect to u is ensured beforehand by an appropriate choice of $\varphi(t, y)$. Indeed, φ is chosen so that at every point (t, y) there exists an angle $\omega \in (1, \dot{\bar{y}})$.

The specific features of each method automatically define the range of problems to which they are best applied: the Hamilton—Jacobi—Bellman method is little sensitive (at least in principle) to constraints on the control function, while being extremely sensitive to constraints imposed on the phase coordinates, the constraints on the vector y included. For the new formalism, on the other hand, the constraints on the phase coordinates are of no significance.

In virtue of the particular construction used, the method of multiple maxima is applicable only to degenerate problems (although not necessarily to branching control). This is a typical method for solving degenerate problems.

The above discussion illustrates the efficiency of the method. Thus, instead of solving a boundary-value problem for a system of differential equations, whose solution is far from covering all the possible control programs (Lagrange's method), or solving a nonlinear partial differential equation (the Hamilton—Jacobi method), the method of multiple maxima reduces the entire problem to elementary relations. This simplification is made possible by the relaxation of the constraints on $y(t)$, $u(t)$ as a result of degeneracy. The condition of the strict equality of the vectors $(1, \dot{\bar{y}})$ and $(1, \dot{f}(t, \bar{y}, \bar{u}))$ in the (t, y) space is replaced by the weaker condition that the vector $(1, \dot{\bar{y}})$ should belong to the angle $\omega(t, \bar{y}(t))$.

We will now consider the generalization of the method of multiple maxima to some common multidimensional $(n>1)$ variational problems.

3. 5. 2. Systems with linear control

Consider the following particular case of the general variational problem (Chapter I). The functional I is given by

$$I = \int_{t_0}^{t_1} f^0(t,y,v)\,dt + F(y_0,y_1), \qquad (3.43)$$

the differential equations (1.3) have the form

$$\dot{y} = f(t,\ y,\ u) \equiv g(t,\ y,\ v) + h(t,\ y)w, \qquad (3.44)$$

where

$$u = (v,\ w),\quad v = (v^1,\ v^2,\ \ldots,\ v^{r-1}),$$

and w is a scalar control function.

The set $V_u(t,\ y)$ is defined by the conditions

$$v \in V_v(t,\ y),\ w \in V_w(t,\ y); \qquad (3.45)$$

$$y \in V_y(t),\ t \in [t_0,\ t_1], \qquad (3.46)$$

where $V_v(t,\ y)$ is a given set of points in the $(r-1)$-dimensional vector space, V_w is the segment $[w_1(t,\ y),\ w_2(t,\ y)]$.

The vector functions $g(t,\ y,\ v) = (g^1,\ g^2,\ \ldots,\ g^n)$ and $h(t,\ y) = (h^1,\ h^2,\ \ldots,\ h^n)$ are continuous and differentiable for all $t \in [t_0,\ t_1]$; $y \in V_y(t),\ v \in V_v(t,\ y)$.

Problems of this kind constitute a special class in variational calculus, since the standard classical necessary and sufficient conditions do not provide a solution for the local minimum of the functional in this case. Thus, Weierstrass's and Legendre's strengthened conditions /3/ a priori do not hold. Jacobi's condition is meaningless for these problems.

To solve the problem, we will apply Theorem 1.1. We have

$$R(t,\ y,\ v,\ w) = \varphi_y(g(t,\ y,\ v) + h(t,\ y)w) - f^0(t,\ y,\ v) + \varphi_t. \qquad (3.47)$$

The function φ being arbitrary to a degree, we define it so that for any admissible t and y,

$$\varphi_y h(t,\ y) = 0. \qquad (3.48)$$

Given this φ, the function R is independent of w.

Condition (3.48) is a partial differential equation for the function φ. The general solution of this equation is an arbitrary continuous and differentiable function

$$\varphi = \varphi_1(t,\eta),$$

where

$$\eta = \eta(t,y); \qquad\qquad\qquad (3.49)$$

$\eta(t,\ y) = (\eta^1,\ \eta^2,\ \ldots,\ \eta^{n-1})$ is a set of the $(n-1)$ independent first integrals of the system of differential equations

$$\frac{dy}{d\tau} = h(t,y). \qquad\qquad\qquad (3.50)$$

We choose $t,\ y$ as the new arguments of the function φ_1. This gives

$$R = R_1(t,\eta,y,v) = \varphi_{1\eta}\left(\sum_{i=1}^{n} \eta_{y^i}g^i(t,y,v) + \eta_t\right) - f^0(t,y,v) + \varphi_{1t}. \qquad (3.51)$$

The problem now reduces to finding the maximum of R_1 for every fixed $t \in (t_0,\ t_1)$ on the set of points $(\eta,\ y,\ v,\ w)$ in the $(n+r+1)$-dimensional set, which satisfy the conditions (3.45), (3.49) and the condition $y \in V_y(t)$.

Formally, the R defined by expression (3.51) is a function of the functional (3.43) and the following system of constraints:

$$\dot{\eta} = \sum_{i=1}^{n} \eta_{yl}g^i(t,y,v) + \eta_t; \qquad\qquad (3.52)$$

$$\eta^j = \eta^j(t,y), \quad j = 1,\ 2,\ldots,\ n-1, \qquad\qquad (3.53)$$

where η^j are phase coordinates, y^l, v^k are control functions.

We will show that in the absence of any constraints on w, the initial problem (we designate it Problem 1) can be reduced, actually as well as formally, to the problem of minimizing the functional (3.43) on the set of elements $(\eta(t),\ y(t),\ v(t),\ w(t))$ satisfying equations (3.51), (3.52) and conditions (3.45), (3.46) (we designate it Problem 2).

Consider a set D_1 of elements $(\eta(t),\ y(t),\ v(t),\ w(t))$ satisfying the following conditions:

1. The vector function $\eta(t) = (\eta^1,\ \eta^2,\ \ldots,\ \eta^{n-1})$ is continuous and piecewise-differentiable on $[t_0,\ t_1]$; η^i are the new phase coordinates.

2. The vector function $(y(t),\ v(t))$ is continuous everywhere on $[t_0,\ t_1]$ with the possible exception of a finite number of points, where it may have discontinuities of the first kind.

3. For every $t \in [t_0,\ t_1]$ the vector $(\eta,\ y,\ v)$ belongs to the set $V_1(t)$ of the $(2n+r)$-dimensional space defined by conditions (3.45), (3.48), and (3.46).

4. The functions $\eta(t),\ y(t),\ v(t)$ satisfy equations (3.51). The set of all control functions $(y(t),\ v(t))$ is designated D_2. Problem 2 is formulated as follows. Find an element $(\eta(t), y(t),\ v(t)) \in D_1$ on which the functional (3.43) attains its minimum value on D_1. For the purposes of Problem 2 it is assumed that the sought element $(\tilde{\eta}(t),\ \tilde{y}(t),\ \tilde{v}(t))$ is contained in D_1. All that follows can be generalized without difficulty to the case when the minimum of I is not attained on D_1 and a minimizing sequence is to be constructed in D_1.

Let us consider the simultaneous equations (3.44), (3.52) for $(y(t),\ v(t),\ w(t)) \in D$. Then, by (3.53), there are only n independent equations among (3.44), (3.52), and the equations in (3.52) are independent of one another. We will now show that $D_2 \supset D$.

Indeed, let D_2' by a subset of elements of D_2 satisfying all the conditions of D, with the exception of equations (3.44). Then $D_2' \supset D$, since system (3.52) contains fewer independent constraints than system (3.52), (3.44), which is equivalent to system (3.44). Hence, D is a subset of D_2, $D_2 \supset D$. Then, in virtue of the particular structure of the functional (3.43), we have the obvious relations

$$\inf_{D_1} I = \inf_{D_2} I \leqslant \inf_{D} I. \qquad (3.54)$$

Suppose that Problem 2 has been solved, say, by further investigation of the function R_1 and the conditions (1.12). Let $(\tilde{y}(t), \tilde{v}(t), \tilde{w}(t))$ minimize the functional (3.43) on D_2, where $\tilde{w}(t)$ may be any piecewise-continuous function on $[t_0, t_1]$, since the functional and system (3.52) are independent of w. We will now show that a function $\bar{w}(t)$ exists such that there is an element $(\bar{y}(t), \bar{v}(t), \bar{w}(t))$ satisfying equations (3.44) which satisfies to any desired accuracy the boundary conditions (3.46) and the constraints (3.45), (3.46) and approximates* to the element $(\tilde{y}(t), \tilde{v}(t), \tilde{w}(t)) \in D_2$.

Indeed, if $\tilde{y}(t)$ is continuous and piecewise-differentiable, with a finite number of points of discontinuity of the derivative, then inserting $\tilde{y}(t)$ in one of the equations in (3.44) (such that the substitution does not give a trivial relation of the form $0 = 0$) and solving the resulting equation for w, we obtain a function $\bar{w}(t)$ which, in virtue of the properties of $\tilde{y}(t)$ and g, h has at most a finite number of discontinuities of the first kind. In this case $(\tilde{y}(t), \tilde{v}(t), \tilde{w}(t)) \in D$ and we may take

$$(\overline{y}(t), \overline{v}(t), \overline{w}(t)) = (\tilde{y}(t), \tilde{v}(t), \overline{w}(t)).$$

If $\tilde{y}(t)$ has a finite number of discontinuities of the first kind, we proceed as follows. Suppose that relations (3.49) are written in the form

$$y = \chi(\eta, \theta, t), \quad \chi = (\chi^1, \chi^2, ..., \chi^n), \qquad (3.55)$$

where $\chi^i(\eta, \theta, t)$ are some constraints which are continuous and differentiable functions of their arguments, θ is some scalar variable.

We know from the theory of ordinary differential equations that such a representation indeed exists (see, e.g., /2/). The substitution of variables reduces (3.52) to the form

$$\dot{\eta} = v(\eta, \theta, v, t), \quad v = (v^1, v^2, ..., v^{n-1}), \qquad (3.56)$$

where $v^j(\eta, \theta, v, t)$ are some continuous functions of their arguments, and (3.44) is reduced to the form

$$\dot{\eta} = v(\eta, \theta, v, t); \qquad (3.57)$$

$$\dot{\theta} = v_1^n(\eta, \theta, v, t) + v_2^n(\eta, \theta, t) w. \qquad (3.58)$$

---

* We say that a continuous vector function $x(t)$ approximates on $[t_0, t_1]$ with an accuracy ε to some piecewise-continuous function $\tilde{x}(t)$ with a finite number of discontinuities of the first kind if $|x(t) - \tilde{x}(t)| < \varepsilon$ everywhere on $[t_0, t_1]$, with the possible exception of the ε-neighborhoods of the discontinuity points of $\tilde{x}(t)$.

The functional (3.43) and the other relations figuring in the problem are also expressed in the new variables. In (3.57), (3.58), θ is a control element, whereas in (3.56) it is a phase coordinate. Let $\widetilde{\theta}(t)$ be the control function $\theta(t)$ corresponding to the solution of Problem 2. In our case, $\widetilde{\theta}(t)$ is some piecewise-continuous function with a finite number of discontinuities of the first kind (Figure 3.2). Evidently,

$$\widetilde{y}(t) = \chi\left(\widetilde{\eta}(t),\ \widetilde{\theta}(t),\ t\right).$$

We partition the segment $[t_0,\ t_1]$ into subintervals Δ_s so that the discontinuities coincide with some of the partition points. Joining the points $(t_s,\ \widetilde{\theta}(t_s))$ by straight segments, we obtain a continuous polygonal line $(t,\ \overline{\theta}_S(t))$ "inscribed" inside the curve $(t,\ \widetilde{\theta}(t))$. Here $s = 1,\ 2, \ldots,\ S$.

Let us find the solution $\overline{\eta}_S(t)$ of (3.56) for $\theta(t) = \theta_S(t)$, $v(t) = \widetilde{v}(t)$ and the initial conditions $\overline{\eta}_S(t) = \widetilde{\eta}(0)$. First we will investigate the behavior of the solution $\overline{\eta}_S(t)$. We see from the construction of $\overline{\theta}_S(t)$ that for any arbitrarily small $\varepsilon > 0$, there exists δ (for $S \to \infty$, max $\Delta_s \to 0$) such that

$$\delta\theta = \left|\widetilde{\theta}(t) - \overline{\theta}_S(t)\right| < \varepsilon$$

everywhere on $[t_0,\ t_1]$, with the possible exception of the ε-neighborhoods of the discontinuity points of $\widetilde{\theta}(t)$. Because of the constraints, we can construct a closed bounded region G in the space of the variables η, θ, t which contains all the admissible values of the variables corresponding to the solution of system (3.52) with the above initial conditions. Since the functions $v^l(\eta,\ \theta,\ t)$ are continuous, the increment

$$\delta v = \left|v\left(\eta,\ \widetilde{\theta}(t),\ \widetilde{v}(t),\ t\right) - v\left(\eta,\ \overline{\theta}_S(t),\ \widetilde{v}(t),\ t\right)\right| \tag{3.59}$$

is of the order ε, i.e., $\delta v < k\varepsilon$ for $\delta\theta < \varepsilon$, where k is some constant. Near the discontinuity points of $\widetilde{\theta}(t)$, $\delta\theta$ and δv are bounded, i.e., $\delta\theta < l$, $\delta v < m$ where l, m are constants. Hence we obtain an estimate for $\delta\eta = \left|\widetilde{\eta}(t) - \overline{\eta}_S(t)\right|$:

$$\delta\eta = \int_0^{t_1} \left|v\left(\widetilde{\theta}(t),\ \widetilde{v}(t),\ t - v(\eta,\ \overline{\theta}_S(t),\ \widetilde{v}(t),\ t\right)\right| dt \leqslant$$

$$\leqslant \int_0^{t_1} \delta v\, dt \leqslant \sum_1^N \left(\int_{\tau_n - \varepsilon}^{\tau_n} \delta v\, dt + \int_{\tau_n}^{\tau_{n+1} - \varepsilon} \delta v\, dt\right) \leqslant N\left[m\varepsilon + k\varepsilon \sum_1^N \tau_{n+1} - \tau_n\right], \tag{3.60}$$

where τ_n, $n = 1,\ 2,\ \ldots,\ N$, are the discontinuity points of $\widetilde{\theta}(t)$.

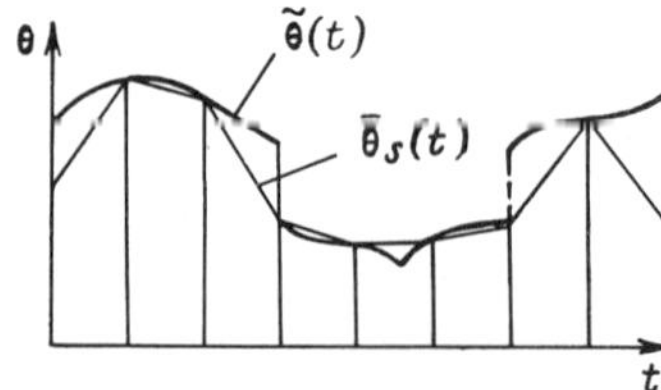

FIGURE 3.2

The same estimate evidently applies to

$$\delta I = |\overline{I}_S - \widetilde{I}|.$$

Hence it follows that for $S \to \infty$ (max $\Delta_S \to 0$), $|\widetilde{\eta}(t) - \overline{\eta}_S(t)| \to 0$ for all $t \in [t_0, t_1]$ (in particular, $\overline{\eta}_S(t_1) \to \widetilde{\eta}(t_1)$).

Inserting $\overline{\theta}_S(t)$, $\overline{\eta}_S(t)$ in (3.55), we obtain

$$\overline{y}_S(t) = \chi(\overline{\eta}_S(t),\, \overline{\theta}_S(t),\, t). \qquad (3.61)$$

The functions $\chi(\eta,\, \theta,\, t)$ being continuous, we obtain

$$\overline{y}_S(t_1) \to \widetilde{y}(t_1)$$

for $S \to \infty$.

Inserting $\overline{\theta}_S(t)$, $\overline{\eta}_S(t)$, $\widetilde{v}(t)$ in (3.58) and solving the resulting expression for w, we obtain some piecewise-continuous function $\overline{w}(t)$ with a finite number of discontinuities of the first kind.

The sequence $(\overline{y}_S(t),\, \widetilde{v}(t),\, \overline{w}_S(t))$ is a minimizing sequence, each of its elements satisfying equations (3.44) and, with an accuracy of ε, the appropriate boundary conditions and constraints; in other words, it belongs to the set D with an accuracy of ε.

We thus arrive at the following general scheme for the solution of the problem. The starting system of equations is replaced by equations (3.52), (3.53), where the phase coordinates are the first integrals of (3.50) and the control functions are $y(t)$, $v(t)$. In virtue of the properties of the first integrals, this system is independent of the control function $\dot{w}(t)$. We thus define a set D_2 of the elements $(y(t),\, v(t))$. We then minimize the functional (3.43) on the set D_2 (in a certain sense, this is a simpler problem, because the new system of differential constraints has a lower order than the original system). The solution obtained in this way either belongs to D or can be approximated with any desired accuracy by a sequence in D which converges to this solution and is at the same time a minimizing sequence of the original problem. In our case, when the control function w is unconstrained, Problems 1 and 2 are in fact equivalent.

In general, when $V_w(t,\, y)$ is bounded, Problems 1 and 2 are not equivalent. The above scheme may be applied in this case also: the only additional step is to check that the function $\overline{w}(t)$ satisfies the constraints. If it does not, the sought solution will contain sections corresponding to the boundary values of w. Our scheme is inapplicable in this case for a rigorous solution of the problem. However, the solution of Problem 2 again proves quite useful: it helps to form a qualitative idea of what the sought solution should be (the discontinuity points of $\widetilde{y}(t)$ generally correspond to the boundary control w) and to obtain a lower-bound estimate of the sought solution (relation (3.54)). The conditions of the optimum principle (1.11), (1.12) are naturally valid in the general case also. The sought minimizing solution in this case is probably a combination of segments corresponding to the interior points of the set V_w (singular sections) and segments corresponding to the boundary of V_w. On singular sections, the function φ is naturally defined as a solution of the partial differential equation (3.48), and after that we proceed to investigate the function R_1.

Singular sections are described by equations (3.52), (3.53). On sections corresponding to the boundary w, φ should be defined proceeding from the specific features of the particular problems being considered (see 3. 5. 4).

Finally, we obtain equations which describe sections of different types and the function φ on these sections. The last step in this solution procedure is the appropriate matching of the different sections and a final verification of conditions (1.11), (1.12).

3. 5. 3. A degenerate quadratic functional

The above method will now be applied to minimize the functional

$$I=\int\limits_{t_0}^{t_1} (a_{ij}(t)y^i y^j + b_i(t)\, y^i w)dt \tag{3.62}$$

with the constraints

$$\dot{y}=M(t)\,y+L(t)\,w, \tag{3.63}$$

$$y(t_0)=y(t_1)=0, \tag{3.64}$$

where

$$y=(y^1,\, y^2,\, \ldots,\, y^n\,);$$

$M(t)$ and $L(t)=(L^1,\, L^2,\, \ldots,\, L^n\,)$ are respectively an $n\times n$ matrix and an n-dimensional vector of the coefficients of linear system (3.63). Summation over repeating indices is implied.

The problem should be reduced to the form (3.43), (3.44), and to this end we will treat it as a problem of minimizing $y^0(t_1)$ for the system

$$\left.\begin{aligned} \dot{y}^0 &= a_{ij}(t)\, y^i y^j + b_i(t)\, y^i w, \\ \dot{y} &= M(t)\, y + L(t)\, w, \end{aligned}\right\} \tag{3.65}$$

with constraints (3.64) and an additional condition $y^0(t_0)=0$.

In this case $f^0(t,\, y,\, v) \equiv 0;\quad F(y_0,\, y_1)\equiv y_1^0;\quad g(t,\, y,\, v) \equiv (a_{ij}y^i y^j,\, M(t)y)$, $h=(b(t)y,\, L(t))$.

Let us find the independent first integrals of (3.65)

$$\frac{dy^0}{d\tau} = b(t)\, y; \tag{3.66}$$

$$\frac{dy}{d\tau} = L(t). \tag{3.67}$$

They have the form

$$\eta^0 = y^0 - \frac{1}{L^n}\, (b_j\eta^j)\, y^n - \frac{1}{2\,(L^n)^2}\, bL\,(y^n)^2; \tag{3.68}$$

$$\eta^j = y^j - \frac{L^j}{L^n}\, y^n, \quad j = 1,2,...,n-1. \tag{3.69}$$

Equations (3.52) take the form

$$\dot{\eta}^0 = a_{ij}y^iy^j - \frac{1}{L^n}\, b_j(M^jy)\, y^n - \frac{1}{L^n}\, (b_j\eta^j)\, M^ny -$$

$$- \left[(b_j\eta^j)\frac{d}{dt}\left(\frac{1}{L^n}\right) - \frac{1}{L^n}\, (b_j\dot{\eta}^j) \right] y^n - \left[\frac{b_jL^j}{L^n} - \frac{1}{(L^n)^2}\, bL \right] \times$$

$$\times (M^ny)\, y^n - \left[b_j\frac{d}{dt}\left(\frac{L^j}{L^n}\right) + \frac{d}{dt}\left(\frac{1}{2(L^n)^2}\, bL\right) \right](y^n)^2; \tag{3.70}$$

$$\dot{\eta}^j = M^jy - \frac{L^j}{L^n}\, M^ny - \left[\frac{d}{dt}\left(\frac{L^j}{L^n}\right) \right] y^n, \tag{3.71}$$

where y^i, η^j satisfy (3.68), (3.69); M^j is the j-th row of the matrix M.

Expressing y^j in terms of η^j and y^n from (3.68), (3.69) and inserting in (3.70) and (3.71), we obtain

$$\dot{\eta}^0 = g_{kl}(t)\,\eta^k\eta^l + g_{kn}(t)\,\eta^ky^n + g_{nn}(t)(y^n)^2, \tag{3.72}$$

$$\dot{\eta}^k = m^k_l(t)\,\eta^l + l^k(t)\,y^n, \tag{3.73}$$

$$k,\, l = 1,2,...,n-1,$$

where $g_{kl}(t),\, g_{kn}(t),\, g_{nn}(t),\, m^k_l(t),\, l^k(t)$ are the coefficients obtained after the substitution and collection of similar terms; $\eta^k(t)$ are the new phase coordinates; $y^n(t)$ are the new control functions.

If $g_{nn}\neq0$, Problem 2 is non-degenerate and can be solved in finite form by the standard methods, e. g., by applying Jacobi's classical condition. The original problem is thus also solved.

A problem of this kind arises in connection with the second variation of the functional if the extremal (the solution of the Lagrange—Euler equations) is degenerate, i. e., if $H_{ww}=0$ along the extremal.

3. 5. 4. An example with constraints on w

Minimize the functional

$$I = \int_0^{t_1} (y^1 - (y^2)^2)\, dt \tag{3.74}$$

with the constraints

$$\dot{y}^1 = y^2w; \tag{3.75}$$

$$\dot{y}^2 = \frac{y^2}{y^1} - w; \tag{3.76}$$

$$y^1 \geqslant 0;\ |w| \leqslant A;\ A > 0; \tag{3.77}$$

$$y^1(0)=y^2(0)=0, \quad y^1(t_1)=y_1^1, \quad y^2(t_1)=y_1^2, \tag{3.78}$$

where t_1 is free.

To solve the problem, we find the first integral of the system

$$\frac{dy^1}{d\tau}=y^2, \quad \frac{dy^2}{d\tau}=-1. \tag{3.79}$$

The first integral is given by

$$\eta=y^1+\frac{1}{2}(y^2)^2. \tag{3.80}$$

Changing over to new variables η, y^1, w in (3.74), (3.75), (3.76), we express $(y^2)^2$ in terms of η and y^1 from (3.80):

$$I=\int_0^{t_1}(3y^1-2\eta)\,dt; \tag{3.81}$$

$$\dot{\eta}=\frac{2(\eta-y^1)}{y^1}; \tag{3.82}$$

$$\dot{y}^1=\sqrt{2(\eta-y^1)}\,w. \tag{3.83}$$

Since the right-hand sides of (3.81) $-$ (3.83) are independent of t and t_1 is free, we can conveniently change over to a new argument η. By (3.77), $\dot\eta\geqslant0$, and this substitution is legitimate. The problem is thus reduced to the equivalent problem of minimizing the functional

$$I_1=\int_{\eta_0}^{\eta_1}f^0(y^1,\eta)\,d\eta\equiv\int_{\eta_0}^{\eta_1}y^1\,\frac{3y^1-2\eta}{\eta-y^1}\,d\eta \tag{3.84}$$

with the constraints

$$(y^1)'=f(y^1,\eta,w)\equiv\frac{y^1}{\sqrt{2(\eta-y^1)}}\,w; \tag{3.85}$$

$$\eta_0=0; \quad \eta_1=y_1^1+\frac{1}{2}(y_1^1)^2; \tag{3.86}$$

$$y^1(\eta_0)=0; \quad y^1(\eta_1)=y_1^1; \quad 0\leqslant y^1\leqslant\eta. \tag{3.87}$$

To solve the problem, we investigate the maximum of the function

$$R=\varphi_{y^1}\,\frac{y^1}{\sqrt{2(\eta-y^1)}}\,w-y^1\,\frac{3y^1-2\eta}{\eta-y^1}+\varphi_\eta. \tag{3.88}$$

R attains a maximum with respect to w under the following conditions:

$$\varphi_{y^1}=0, \text{ any } w \tag{3.89}$$

$$\varphi_{y^1}>0, \quad w=A; \tag{3.90}$$

$$\varphi_{y^1}<0, \quad w=-A. \tag{3.91}$$

In the first case, R takes the form

$$R = -y^1\,\frac{3y^1 - 2\eta}{\eta - y^1} + \varphi_\eta. \tag{3.92}$$

The dependence of R on y^1 for some fixed η is shown in Figure 3.3.

It is readily checked that the maximum R for every fixed η is attained for $y^1(\eta) = \left(1 - \sqrt{\dfrac{1}{3}}\right)\eta$ (see Figure 3.3). The solution of Problem 2 thus has the form

$$\left.\begin{aligned}
\widetilde{y}^1 &= \left(1 - \sqrt{1/3}\right)\eta \quad (0 \leqslant \eta < \eta_1);\\
\widetilde{y}^1 &= y_1^1 \quad (\eta = \eta_1).
\end{aligned}\right\} \tag{3.93}$$

It is readily seen that this solution gives an absolute minimum of the functional (3.84), if the differential constraint is ignored.

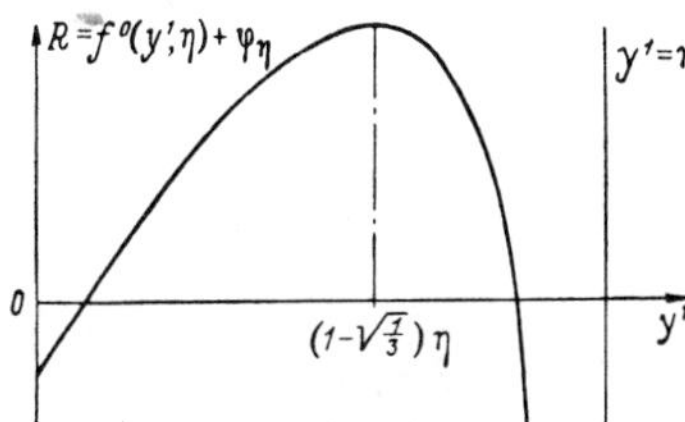

FIGURE 3.3

Inserting (3.93) in (3.85), we see that $w = \infty$ for $\eta = 0$, $\eta = \eta_1$, so that the constraints imposed on w break down. Therefore, on the initial and the final sections, w is boundary control. Using this information, we construct a trail minimizing solution, which consists of the following three sections (Figure 3.4):

I. $0 \leqslant \eta \leqslant \eta_2$, $(\bar{y}(\eta))_\mathrm{I}$ is the solution of equation (3.85) for $w = A$ and the boundary condition $y^1(\eta_0) = 0$.

II. $\eta_2 \leqslant \eta \leqslant \eta_3$, $\left(\bar{y}(\eta)\right)_\mathrm{II} = \widetilde{y}(\eta)$.

III. $\eta_3 \leqslant \eta \leqslant \eta_1$, $\left(\bar{y}(\eta)\right)_\mathrm{III}$ is the solution of equation (3.85) for $w = A$ and the boundary condition $y^1(\eta_1) = y_1^1$.

The function $\varphi(\eta, y^1)$ is defined on I, II in the form

$$\varphi(\eta,\ y^1) = \psi(\eta)\,y^1 + \sigma(\eta)(y^1 - \bar{y}^1(\eta))^2, \tag{3.94}$$

where $\psi(\eta)$, $\sigma(\eta)$ are some continuous and piecewise-differentiable functions of η.

The sufficient conditions of a local maximum of R on I and III have the form

$$\varphi_{y^1} > 0; \tag{3.95}$$

$$P_{y^1} \equiv R_{y^1}(\eta,\ y^1,\ A) \equiv \psi' + H_{y^1} = 0, \tag{3.96}$$

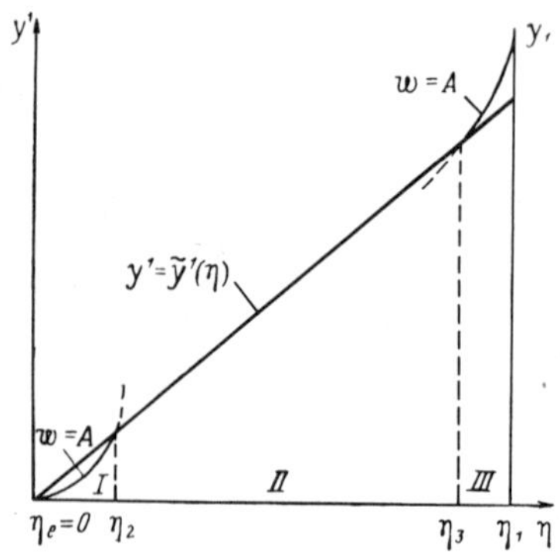

FIGURE 3.4

$$P_{y^1y^1} \equiv R_{y^1y^1}(\eta, y^1, A) \equiv \sigma' + 2\sigma \left(\frac{y^1}{\sqrt{2(\eta - y^1)}} \right)_{y^1} A + H_{y^1y^1} < 0, \qquad (3.97)$$

where

$$H = \psi f(y^1, \eta, A) - f^0(y^1, \eta),$$

$$P(\eta^1) = \sup_{|w| \leqslant A} R(\eta, y^1, w). \qquad (3.98)$$

At the points η_2, η_3 we should have

$$\eta = \eta_2, \ (\psi)_I = 0, \ (\sigma)_I = 0; \qquad (3.99)$$

$$\eta = \eta_3, \ (\psi)_{III} = 0, \ (\sigma)_{III} = 0. \qquad (3.100)$$

Otherwise, the function $\varphi(\eta, y^1)$ will be discontinuous in y^1 at these points (on section II, $\varphi_{y^1} = 0$).

It is readily seen that the solution of equation (3.96) on III with boundary conditions (3.100) is positive.

Indeed, consider the general solution of (3.96)

$$\psi = e^{-F} \int_{\eta_3}^{\eta_1} g(\eta) e^F dy; \ \ F(\eta) = \int_{\eta_3}^{\eta_1} h(\eta) dy,$$

where $h(\eta)$ is a coefficient before ψ in (3.96); $g(\eta) = f^0_{y^1}$.

On this section $f_{0y^1} > 0$ (see Figure 3.3). Hence $\psi(\eta) > 0$ for all $\eta \in (\eta_3, \eta_1)$.

A similar situation is observed on section I. The only difference is that the boundary condition here is defined for the right end point and $f^0_{y^1} < 0$, since $y^1 < \bar{y}_1$. Here again $\psi(\eta) > 0$ everywhere on $[\eta_0, \eta_2]$. Since φ_η is continuous, we conclude that there exists a neighborhood of the curve $\bar{y}^{(1)}(\eta)$ where

$$\varphi_\eta \geqslant 0. \qquad\qquad (3.101)$$

Consider condition (3.97).   This inequality may be replaced by an equation with a positive right-hand side

$$\sigma' + 2\sigma f_{1y^1} A + H_{y^1 y^1} = \varepsilon(\eta), \qquad\qquad (3.102)$$

where $\varepsilon(\eta) > 0$ is some piecewise-continuous function of η .

Equation (3.102) with boundary conditions (3.100) is a priori solvable on I and III (a Cauchy problem for a linear differential equation).   Hence it follows that there exists a function $\varphi(\eta, y^1)$ such that the solution constructed on I, III corresponds at least to a local maximum of $P(y^1, \eta)$, and the solution on II corresponds to an absolute maximum of $R(\eta, y^1, v)$. We thus see that the solution constructed above ensures at least a strong local minimum of the functional in our problem.

In this example, Problems 1 and 2 are not equivalent because of constraints (3.77) imposed on the control function w.   As a result, we did not change over to Problem 2 at the very beginning.   However, we did change over to new phase coordinates (y^1, η) and the solution of the problem was thus considerably simplified.

Remarks

1.   Problem 2 (the function R_1) may turn out to have the same singularities as the original problem (the function R).   The method of multiple maxima then can be applied repeatedly to analyze Problem 2.

2.   A practical shortcoming of the above method is that it requires determination of the first integrals of some system of ordinary differential equations (3.50).   In applied problems and, in particular, in most problems of flight dynamics, equations (3.50) are sufficiently simple and their first integrals are obtained without difficulty.

3.   The above method can be generalized without any changes to the case when the vector function h depends on the control function v, but so that for fixed t and y the vector h maintains a constant direction when v is varied.   In other words, the unit vector function $h(t, y, v)/|h(t, y, v)|$ is independent of v, and no constraints are imposed on w.

We write $h(t, y, u)w$ in the form $h_1(t, y)w_1$, where $h_1 = \dfrac{h}{|h|}$, $w_1 = |h|w$, and w_1 is used as the new control function.   The problem is thus reduced to the one considered above.

3.5.5.   Systems with several unconstrained control functions

Let the functional (1.1) and system (1.3) have the form

$$I = \int_{t_0}^{t_1} f^0(t,y)\,dt + F(y_0, y_1), \qquad\qquad (3.103)$$

$$\dot{y}=g(t,y)+\sum_{l=1}^{k} h_l(t,y)\,u^l,\tag{3.104}$$

where u^l, $l=1, 2, \ldots, k$, are the components of the control vector u. The set V_u coincides with the space U. The vector functions h_l are continuous and differentiable for every $t \in (t_0, t_1)$, $y \in V_y$, and the functions g, f^0 are moreover bounded for bounded t, y, u.

Although the problem can be solved by the method of the previous subsection, using several recursive reductions to Problem 2, we will describe a different approach which a priori enables us to indicate the conditions to be satisfied by the vector functions $h_l(t, y)$ so that the function

$$R=\varphi_y\left(g(t,y)+\sum_{l=1}^{k} h_l(t,y)u^l\right)-f^0(t,y)+\varphi_t\tag{3.105}$$

for every $t \in (t_0, t_1)$ attains a maximum on

$$\bar{y}(t) \in V_y(t),\ \bar{u}(t) \in V_u,\ |\bar{u}|\neq\infty.\tag{3.105a}$$

For simplicity, we assume $V_y(t)$ to be an open bounded region. We write R in the form

$$R=A(t, y)u+B(t, y),\tag{3.106}$$

where $A(t, y)$ is a vector with components $\varphi_y h_l(t, y)$, $l=1, 2, \ldots, k$,

$$B(t, y)=\varphi_y g(t, y)+\varphi_t-f^0(t, y).$$

First we prove the following lemma.

Lemma. For the function (3.106) to attain a maximum on (3.105a) for a fixed t, it is necessary and sufficient that

1) $A(t,y)\equiv 0,$ $\qquad\qquad\qquad\qquad\qquad\qquad\qquad\qquad$ (3.107)

2) $B(t,\bar{y})=\max\limits_{y\in V_y(t)} B(t,y).$ $\qquad\qquad\qquad\qquad\qquad\qquad$ (3.108)

Proof. **Necessity.** We write the vector u in the form

$$u=\varrho v,$$

where $v=|u|$; ϱ is a unit vector $\dfrac{u}{|u|}$.

The inequality

$$R(t, y, u)-R(t, \bar{y}(t), \bar{u}(t))\leqslant 0$$

is satisfied for all $y \in V_y(t)$, $u \in U$, if and only if

$$\max_{\varrho} R-R(t,\bar{y},\bar{u})=\max_{\varrho}\{A(t,y)\varrho v+B(t,y)\}-$$
$$-R(t,\bar{y},\bar{u})=|A(t,y)|\,v+B(t,y)-R(t,\bar{y},\bar{u})\leqslant 0\tag{3.109}$$

for all $y \in V_y(t)$, $v \in [0, +\infty]$.

The maximum with respect to ϱ is accounted for as follows. The scalar product $A\varrho$ is the projection of the vector A on the direction of ϱ. It attains a maximum, equal to the magnitude of A, if ϱ points along the vector A.

Suppose that (3.107) is not satisfied and $A(t, y^*(t)) \neq 0$ for some $y = y^*(t) \in V_y(t)$. Then, since the first term in (3.109) is positive and the terms $B(t, y^*(t))$ and $R(t, \bar{y}, \bar{u})$ are finite, we can find such v that inequality (3.109) is not true.

This proves the validity of (3.107) and hence of (3.108), since from (3.107) we have

$$R(t, y, u) \equiv B(t, y).$$

The **sufficiency** of (3.107) and (3.108) is self-evident. Q. E. D.

Condition (3.107) is a system of linear homogeneous partial differential equations for the function φ:

$$L_l(\varphi) = \sum_{i=1}^{n} \varphi_{y^i} \cdot h_l^i(t, y) = 0. \tag{3.110}$$

In general, such systems are incompatible. A necessary condition of their compatibility is that the so-called Poisson brackets for the functions $L_l(t, y, \varphi_y)$ vanish identically (see, e. g., /9/, p. 365):

$$(L_l, L_m) = \sum_{j=1}^{n} \left(\sum_{i=1}^{n} h_l^i \frac{\partial h_m^j}{\partial y^i} - \sum_{i=1}^{n} h_m^i \frac{\partial h_l^j}{\partial y^i} \right) \varphi_{y^j} = 0. \tag{3.111}$$

Conditions (3.111) constitute a new system of linear homogeneous partial differential equations for φ. Seeing that

$$(L_l, L_m) = (L_m, L_l),$$

we conclude that the number of these equations is $\dfrac{k(k-1)}{2}$.

These equations should be added to (3.110), eliminating from the combined set identities and equations which are linear combinations of other equations.

The subsequent stages of the procedure are described in /9/, and no details are given here.

Conditions (3.111) will be used in the next subsection.

3. 5. 6.   Minimum of the second variation of the functional in case of degenerate control

Consider the system of differential equations (3.11), (3.12) which describe the case of degenerate control of degeneracy k. The integrand in the functional (1.1) may be written in the form /2/

$$F^0 = \gamma^\beta(t)\, f^0(t,y,u_\beta), \tag{3.112}$$

$$\beta = 1,\ldots,k+1.$$

Equations (3.11) may be treated as an ordinary system of differential equations with linear control functions γ^β. Solving (3.12) for γ^1, say, and inserting the result in (3.11), we find

$$\overset{v}{y} = F = f(t,y,u_1) + \gamma^\beta\,(f(t,y,u_\beta) - f(t,y,u_1)); \tag{3.113}$$

$$F^0 = f^0(t,y,u_1) + \gamma^\beta\,(f^0(t,y,u_\beta) - f^0(t,y,u_1)), \tag{3.114}$$

$$\beta = 2,\,3,\ldots,\,k+1,$$

where γ^β are now independent and satisfy only the following constraints:

$$\gamma^\beta \geqslant 0; \quad \sum_{\beta=2}^{k+1} \gamma^\beta \leqslant 1, \tag{3.115}$$

Suppose that a degenerate control is observed on some section, satisfying (3.27) with the zero closeness function $y(t) = \bar{y}(t)$ and the basis control functions

$$u = \bar{u}_\beta\,(t),\ \beta = 2,\,3,\,\ldots,\,k+1,$$

$$\gamma^\beta = \overline{\gamma^\beta}\,(t),\ \beta = 2,\,3,\,\ldots,\,k+1,$$

where $\overline{\gamma^\beta}\,(t)$ satisfy the strict inequalities in (3.115):

$$\overline{\gamma^\beta}\,(t) > 0; \quad \sum_{\beta=2}^{k+1} \overline{\gamma^\beta}\,(t) < 1. \tag{3.115a}$$

Adding small increments $\delta\gamma^\beta$ to the functions $\gamma^\beta\,(t)$ such that (3.115a) are still satisfied for the incremented functions $(\overline{\gamma^\beta} + \delta\gamma^\beta)$ with unaltered $\bar{u}_1(t),\ \bar{u}_\beta(t),\ \beta = 2,\,3,\,\ldots,\,k+1,$ and fixed boundary conditions $y(\tau_1) = \bar{y}(\tau_1)$, we obtain from equations (3.113), (3.114) the corresponding increments of the zero closeness function, the integrand function, and hence the functional relative to the values on $(\bar{y}(t),\ \overline{\gamma^\beta}\,(t),\ \bar{u}(t),\ \bar{u}_\beta\,(t))$. For sufficiently small increments $\delta\gamma^\beta$, we can retain only the linear components of the increments $y(t) - \bar{y}(t)$, which are described by a linear system of variational equations

$$\dot{z}^i = \overline{F}_{yj}\,(t)\,z^i + \overline{F}_{\gamma^\beta}\,(t)\,v^\beta, \tag{3.116}$$

where

$$z^i = y^i - \bar{y}^i; \quad v^\beta = \gamma^\beta - \overline{\gamma^\beta},$$

and investigate the sum of the first and the second variation of the functional to within terms of higher order.

It is readily seen that the first variation of the functions is zero by (3.27) and it thus remains to check that the second variation is non-negative, this being the necessary condition for the minimum of the functional.

Let $\mathcal{H}$ be the Hamiltonian of (3.113), (3.114):

$$\mathcal{H}=\psi F-F^0=H(t,\,\psi,\,y,\,u_1)-\gamma^\beta\left(H(t,\,y,\,\psi,\,u_\beta)-H(t,\,y,\,\psi,\,u_1)\right), \qquad (3.117)$$

where

$$H(t,\,y,\,\psi,\,u)=\psi f(t,\,y,\,u)-f^0(t,\,y,\,u)$$

is the Hamiltonian of (1.3). The second variation of the functional is then written in the form (see, e. g., /11/)

$$\delta^2 I=-\int_{\tau_1}^{\tau_1}\left(\overline{\mathcal{H}}_{y^iy^j}(t)\,z^iz^j+\overline{\mathcal{H}}_{y^i\gamma^\beta(t)}z^iv^\beta\right)dt, \qquad (3.118)$$
$$i,\,j=1,\,2,\,\ldots,\,n; \quad \beta=2,\,3,\,\ldots,\,k+1.$$

The superior bar in (3.116) and (3.118) identifies the first derivatives of the vector function F and the function $\mathcal{H}$. Summation over repeating indices is implied.

The problem reduces to investigating the minimum of the functional (3.118) under conditions (3.116) and the boundary conditions

$$z(\tau_1)=z(\tau_2)=0.$$

No restrictions are imposed on the variables z^i, v^β.

This problem thus corresponds to the case considered in the previous subsection. To finally reduce this problem to the form (3.103), (3.104), we will formulate it as Mayer's problem, i. e., the problem of minimizing the final value of z for zero initial condition, supplementing (3.116) with the equation

$$\dot{z}^0=\overline{\mathcal{H}}_{y^iy^j}(t)\,z^iz^j+\overline{\mathcal{H}}_{y^i\gamma^\beta}(t)\,z^iv^\beta. \qquad (3.119)$$

Equations (3.110) now take the form

$$L_\beta(\varphi)=\varphi_{z^0}\overline{\mathcal{H}}_{y^i\gamma^\beta}(t)\,z^j+\varphi_{z^i}\overline{F}_{\gamma^\beta}^i(t)=0. \qquad (3.120)$$

Here

$$h_\beta^0=\mathcal{H}_{y^q\gamma^\beta}z^q;\quad h_\beta^i=F_{\gamma^\beta}^i(t)=f^i\Big|_{u_1}^{u_\beta}.$$

Note that only h_β^0 depends on the phase coordinates; the remaining h_β^i are independent of z, so that their derivatives with respect to z^q in the Poisson brackets all vanish.

Inserting the expressions for h_β^0, h_β^i in (3.111), we find

$$(L_\beta,\,L_\alpha)=\left(\overline{f}\Big|_{u_1}^{u_\beta}\overline{H}_y\Big|_{u_1}^{u_\alpha}-f\Big|_{u_1}^{u_\alpha}\cdot\overline{H}_y\Big|_{u_1}^{u_\beta}\right)\varphi_{z^0}. \qquad (3.121)$$

These combinations should vanish identically in virtue of (3.111). But $\varphi_{z^0}\neq0$. Otherwise, condition (1.12) of Theorem 1.1 is not satisfied. We should therefore have

$$\overline{f}\Big|_{u_1}^{u_\beta}\overline{H}_y\Big|_{u_1}^{u_\alpha}-\overline{f}\Big|_{u_1}^{u_\alpha}\overline{H}_y\Big|_{u_1}^{u_\beta}=0. \qquad (3.122)$$

Both terms on the right are scalar products of the vectors f and H_y.

Now turning to equations (3.29) and (3.12), we express γ^1 from (3.12) in terms of the other γ^β:

$$\gamma^1 = 1 - \sum_{\beta=2}^{k+1} \gamma^\beta.$$

Inserting γ^1 in (3.29) and collecting similar terms, we obtain

$$\gamma^\beta \left[\overline{H}_y \Big|_{u_1}^{u_\alpha} f \Big|_{u_1}^{u_\beta} - H_y \Big|_{u_1}^{u_\beta} f \Big|_{u_1}^{u_\alpha} \right] = - H_t \Big|_{u_1}^{u_\alpha} - \left[H_y \Big|_{u_1}^{u_\alpha} + f(u_1) - H_y(u_1) f \Big|_{u_1}^{u_\alpha} \right], \tag{3.123}$$

$$\alpha, \ \beta = 2, 3, \ldots, k+1,$$

and the coefficients of these equations coincide with the left-hand sides of (3.122) and should therefore vanish. The number of these equalities is $\dfrac{k(k-1)}{2}$, and, as we saw in § 3.2, in virtue of (3.123) they lead to k further equalities

$$H_t \Big|_{u_1}^{u_\alpha} + \left[H_y \Big|_{u_1}^{u_\alpha} f(u_1) - H_y(u_1) f \Big|_{u_1}^{u_\alpha} \right] = 0. \tag{3.124}$$

Equalities (3.122) were obtained as the necessary conditions of a maximum of R and a minimum of Φ when minimizing the second variation in degenerate control problems. They were originally derived in /1/ as the necessary conditions for optimal degenerate control.

Note that once conditions (3.122) have been verified, the problem of minimizing the second variation can be solved to completion by successively changing over to Problem 2.

3.5.7. A more general problem

Consider the same problem as in 3.5.2, using differential equations of a slightly more general form:

$$\dot{y} = g(t, y, v) + h(t, y, w), \tag{3.125}$$

where $g(t, y, v)$ has the same properties as in 3.5.2, and $h(t, y, w)$ and the boundaries $w_1(t, y)$ and $w_2(t, y)$ from (3.45) are such that the functions $h(t, y, w_1(t, y))$ and $h(t, y, w_2(t, y))$ are continuous and continuously differentiable with respect to t and y for all $t \in (t_0, t_1)$, $y \in V_y(t)$.

The function R for this problem has the form

$$R(t, y, v, w) = \varphi_y g(t, y, v) + \varphi_y h(t, y, w) - f^0(t, y, v) + \varphi_t. \tag{3.126}$$

We choose $\varphi(t, y)$ so that for all $t \in (t_0, t_1)$, $y \in V_y(t)$ the function R attains its absolute maximum on two values of w, $w_1(t, y)$ and $w_2(t, y)$,

$$R(t, y, v, w_1(t, y)) = R(t, y, v, w_2(t, y)) = \sup_{w \in V_w(t, y)} R(t, y, v, w). \tag{3.127}$$

We see from (3.127) that only the term

$$\varphi_y \cdot h(t, y, w)$$

depends on w.

Condition (3.127) reduces to a certain combination of inequality constraints

$$\varphi_y \in \Omega(t, y), \tag{3.128}$$

where $\Omega(t, y)$ is a set in the space of vectors φ_y, and the condition

$$\varphi_y(h(t, y, w_1(t, y)) - h(t, y, w_2(t, y)) = 0, \tag{3.129}$$

which is a linear partial differential equation for the function φ. Its general solution is an arbitrary continuous and differentiable function of the form

$$\varphi = \varphi(t, \eta),$$

where $\eta(t, y) = (\eta^1, \eta^2, \ldots, \eta^{n-1})$ is the set of independent first integrals of the system of ordinary differential equations

$$\frac{dy}{d\tau} = h(t, y, w_1(t, y)) - h(t, y, w_2(t, y)), \tag{3.130}$$

with t treated as a parameter.

Let $R_1 = \sup_{w \in V_w} R$. Then

$$R_1 = \varphi_{1\eta} \sum_1^n \eta_{y^i} g^i(t, y, v) + \eta_t.$$

We will further investigate the maximum of the function R_1 for every $t \in (t_0, t_1)$ on the set of points (η, y, v) satisfying conditions (3.45), (3.46) and the additional conditions

$$\eta = \eta(t, y). \tag{3.131}$$

It is further assumed that the function $\varphi(t, y)$ satisfies (3.128).

Suppose that such a function $\varphi(t, y)$ exists and a solution $(\tilde{\eta}(t), \tilde{y}(t), \tilde{v}(t))$ has been found which under the given conditions ensures an absolute maximum of the function R_1 for every $t \in [t_0, t_1]$; moreover, conditions (1.12) and (3.128) are also satisfied. Suppose that this solution satisfies the following system of differential equations:

$$\dot{y} = v_1(g(t, y, v) + h(t, y, w_1)) + v_2(g(t, y, v) +$$
$$+ h(t, y, w_2)) = g(t, y, v) + v_1 h(t, y, w_1) + v_2 h(t, y, w_2), \tag{3.132}$$

where $v_1(t)$, $v_2(t)$ are some piecewise-continuous functions,

$$v_1(t) + v_2(t) = 1.$$

If substituting the solution $(\tilde{y}(t), \tilde{v}(t))$ in this system we find that v_1, v_2 satisfy the conditions

$$v_{1,2}(t) \geqslant 0, \tag{3.133}$$

then $(t, \tilde{y}) \in \omega$ and the solution constitutes a zero closeness curve of a sliding control with the basis control functions $w_1(t, y)$, $w_2(t, y)$ and the basis vectors

$$a_1 = [1, \; f(t, \; y, \; v, \; w_1)]; \\ a_2 = [1, \; f(t, \; y, \; v, \; w_2)]. \quad\quad (3.134)$$

Thus, if all the above conditions are satisfied, our solution defines the sought minimizing sequence and the problem is solved. If no function $\varphi(t, \; y)$ exists and no solution $(\tilde{\eta}(t), \tilde{y}(t), \tilde{v}(t))$ satisfying all the above conditions, we conclude that the sought optimal control may consist of sections of different types — boundary sections, Euler sections, sliding control sections. On sliding control sections, the function $\varphi(t, \; y)$ may be defined by the method that we described.

As in a problem with linear control functions w, we can change over to Problem 2 as follows.

The initial equations (3.125) are supplemented with the expressions for the the total derivatives of $\eta(t, \; y)$ with respect to t:

$$\dot{\eta} = \sum_{1}^{n} \eta_{y^l} \, (g^l(t, \; y, \; v) + h^l(t, \; y, \; w)) + \eta_t. \quad\quad (3.135)$$

Equations (3.125) and (3.135) considered jointly constitute a system of $(2n-1)$ equations where, in virtue of (1.131), only n equations are independent and which is obviously equivalent to the initial system (3.125).

We further consider the problem of minimizing the functional (3.43) on the set L of the elements $(\eta(t), \; y(t), \; v(t), \; w(t))$ which only satisfy equations (3.135) and (3.131), where $\eta^i(t)$ are new phase coordinates and $y, \; v, \; w$ are new control functions.

The remaining conditions (3.45)−(3.46) do not change. The set of new control functions $(y(t), \; v(t), \; w(t))$ is designated D_2. Repeating the same arguments as in 3.5.2, we readily see that $D_2 \supset D$ and

$$\inf_{D_2} I = \inf_{D_1} I \leqslant \inf_{D} I.$$

Suppose that Problem 2 has been solved. The corresponding optimal control $\tilde{w}(t)$ may take on values corresponding either to one of the boundaries $w_1, \; w_2$ of V_w or to its interior points. In the former case, both w_1 and w_2 are optimal values, $\tilde{w} = w_{1,2}$. Indeed, the functions $\eta^i(t, \; y)$ are solutions of the partial differential equations (3.129) (a property of the first integrals of (3.130)). Hence it follows that the right-hand sides of (3.135) are not affected when w_1 is replaced with w_2 and vice versa, so that $\dot{\eta}$ remains unchanged. The degeneracy in this case is $k=1$ and the functions $\tilde{y}(t), \; \tilde{v}(t), \; \tilde{w}(t)$ should only satisfy equations (3.135) and condition (3.45); $\tilde{y}(t), \; \tilde{v}(t), \; \tilde{w}(t)$ can be approximated to any accuracy by constructing an appropriate sequence in D (sliding control). The solution of Problem 2 can be constructed in D under certain boundary conditions. In the second case, it is not the properties of $\eta(t, \; y)$ that cause non-uniqueness of $\tilde{w}(t)$. As a rule, $\tilde{w}(t)$ is unique in this case and, when substituted with $\tilde{v}(t)$ in the original system (3.125), it defines under the appropriate initial conditions a unique solution $y^*(t)$, which does not necessarily coincide with the $\tilde{y}(t)$ derived from other conditions. In this case, the solution of Problem 2 in general cannot be constructed in D.

Bibliography for Chapter III

1. Vapnyarskii, I.B. Teorema sushchestvovaniya optimal'nogo
   upravleniya v zadache Bol'tsa, nekotorye ee prilozheniya i neob-
   khodimye usloviya optimal'nosti skol'zyashchikh i osobykh
   rezhimov (The Theorem of Existence of the Control in Bolza's
   Problem, Some of Its Applications, and the Necessary Conditions
   of Optimality of Sliding and Singular Controls). — Zhurnal Vychislitel'noi Matematiki i Matematicheskoi Fiziki, No. 2. 1967.
2. Gamkrelidze, R.V. O skol'zyashchikh optimal'nykh rezhimakh
   (Sliding Optimal Controls). — Doklady AN SSSR, 143, No. 6. 1962.
3. Ioffe, A.D. Preobrazovaniya variatsionnykh zadach (Transformation of Variational Problems). — Izvestiya AN SSSR, tech.
   cybernetics, Nos. 3, 4. 1967.
4. Krotov, V.F. O razryvnykh rezheniyakh v variatsionnykh zadachakh
   (Discontinuous Solutions of Variational Problems). — Izvestiya
   Vuzov, math. ser., No. 2. 1961.
5. Krotov, V.F. Ob absolyutnom minimume funktsionalov na
   sovokupnosti funktsii s ogranichennoi proizvodnoi (The Absolute
   Minimum of Functionals on the Set of Functions with a Bounded
   Derivative). — Doklady AN SSSR, 140, No. 3. 1961.
6. Krotov, V.F. Razryvnye resheniya variatsionnykh zadach
   (Discontinuous Solutions of Variational Problems). — Izv. Vuzov,
   math. ser., No. 5. 1960.
7. Krotov, V.F. and M.Ya. Brovman. Ekstremal'nye protsessy
   plasticheskogo deformirovaniya metallov (Extremum Processes
   of Plastic Deformation of Metals). — Izvestiya AN SSSR, mechanics
   and machine building series, No. 3. 1962.
8. Lavranet'ev, M.A. and L.A. Lyusternik. Osnovy variatsionnogo ischisleniya (Elements of Variational Calculus), Vol. 1, Parts
   1 and 2. — ONTI. 1932.
9. Smirnov, V.I. Kurs vysshei matematiki (A Course in Higher
   Mathematics), Vol. IV. — Fizmatgiz. 1958.
10. Stepanov, V.V. Kurs differentsial'nykh uravnenii (A Course in
    Differential Equations). — Gostekhizdat. 1945.
11. Troitskii, V.A. O variatsionnykh zadachakh optimizatsii
    protsessov upravleniya (Variational Optimization Problems for
    Control Processes). — Prikladnaya Matematika i Mekhanika,
    26, No. 2. 1962.

Chapter IV

SOME PROBLEMS OF POWERED
FLIGHT OPTIMIZATION

§ 4.1. VERTICAL ASCENT OF A ROCKET IN VACUUM

The present section deals with a relatively simple problem of rocket dynamics, namely that of reaching extreme (maximum) altitude in one-dimensional vertical motion in vacuum (airless space). Constant gravitational acceleration is assumed. This problem may be treated as a simplified model of the general optimization problem of rocket maneuvers in a homogeneous field, which is solved in the next section. The method of solution and the specific properties of the solution of the general problem are conveniently illustrated in this simple case.

The equations of the vertical motion of a rocket with controlled thrust may be written in the form

$$h' = -\frac{1}{\beta} V; \qquad (4.1)$$

$$V' = -\frac{c}{m} + \frac{1}{\beta} g; \qquad (4.2)$$

$$t' = -\frac{1}{\beta}, \qquad (4.3)$$

where h is the altitude reckoned from the surface of the planet, V is the velocity, t is time, m is the mass, c is the nozzle velocity, g is the gravitational acceleration (assumed constant), β is the per-second rate of fuel consumption (the control element), $0 \leqslant \beta \leqslant \beta_{max}$. The independent variable in these equations is the current mass (a non-increasing function of time).

The following boundary conditions are used:

$$h_0 = 0, \ V_0 = 0, \ t_0 = 0 \qquad (4.4)$$

for $m = m_0$ (the rocket starts from rest); the values of v_1 and t_1 for $m = m_1$ are not given.

Our problem is to find the motion reaching the maximum altitude at the end of the powered flight (i.e., minimize $(-h_1)$).

The solution of the problem is divided into two stages: first we solve the problem for any fixed t_1 such that

$$(t_1 - t_0) \geqslant (m_0 - m_1)\, 1/\beta_{max},$$

and then carry out optimization with respect to t_1, which yields the final solution.

The functions R and Φ of our problem are written in the form

$$R = \varphi_h \left(-\frac{1}{\beta} V \right) + \varphi_V \left(-\frac{c}{m} + \frac{1}{\beta} g \right) - \varphi_t \frac{1}{\beta} + \varphi_m; \qquad (4.5)$$

$$\Phi = -h_1 + \varphi(m_1,\ h_1,\ V_1,\ t_1). \qquad (4.6)$$

Note that $\dfrac{1}{\beta}$ may be treated as a linear control function and equations $(4.1) - (4.3)$ have the same form as the equations treated in 3. 5. 2.

We will use the method of multiple maxima. Solving the partial differential equation

$$-\varphi_h V + \varphi_V g - \varphi_t = 0, \qquad (4.7)$$

we obtain the first integrals $\xi,\ \eta$ of the characteristic system

$$\frac{dh}{dt} = V, \quad \frac{dV}{dt} = -g, \qquad (4.8)$$

namely

$$\xi = V + gt, \quad \eta = h - Vt - \frac{1}{2} gt^2. \qquad (4.9)$$

The general solution of (4.7) is an arbitrary continuous differentiable function $\varphi_1(\xi,\ \eta,\ m)$. Inserting this function in (4.5), (4.6) and changing over to new variables ξ and η, we find

$$R_1 = (\varphi_{1\xi} - \varphi_{1\eta} t) \left(-\frac{c}{m} \right) + \varphi_m; \qquad (4.9a)$$

$$\Phi_1 = -\eta_1 - \frac{1}{2} gt_1^2 - (\xi_1 - gt_1)\, t_1 + \varphi_1(m_1,\ \xi_1,\ \eta_1). \qquad (4.10)$$

The function φ_1 is given in the form

$$\varphi_1 = \psi_{(\xi)} \xi + \psi_{(\eta)} \eta, \qquad (4.11)$$

where $\psi_{(\xi)},\ \psi_{(\eta)}$ are some constants. Then

$$\varphi_{1m} = 0; \quad \varphi_{1\xi} = \psi_{(\xi)}; \quad \varphi_{1\eta} = \psi_{(\eta)}, \qquad (4.12)$$

and (4.10) takes the form

$$\Phi_1 = -\eta_1 - \xi_1 t_1 + \psi_{(\xi)} \xi_1 + \psi_{(\eta)} \eta_1 + \frac{1}{2} gt_1^2. \qquad (4.13)$$

Taking $\psi_{(\xi)} = t_1,\ \psi_{(\eta)} = 1$, we find

$$\Phi_1 = \frac{1}{2} gt_1^2 = \text{const} = \min_{\xi_1, \eta_1} \Phi_1 \qquad (4.14)$$

for any $\xi_1,\ \eta_1$.   The function R_1 takes the final form

$$R_1 = -(t_1 - t)\frac{c}{m}\ .\tag{4.15}$$

Let us investigate the minimum of this function* for any fixed $m \in (m_1,\ m_0)$. The minimum is clearly obtained at the lower limit value of t, i.e., $t_L(m)$.

The extreme values of t over the interval $(m_0,\ m_1)$, i.e., $t_L(m)$ and $t_U(m)$, are obtained from equation (4.3) with the appropriate boundary conditions and the constraints on β. The lower limit $t_L(m)$ is the solution of equation (4.3) for $\beta = \beta_{\max}$ which passes through the point $(m_0,\ t_0)$, and $t_U(m)$ is the same solution passing through the point $(m_1,\ t_1)$ (Figure 4.1).

Indeed, no solution of (4.3) passing through $(m_0,\ t_0)$ may lie below $t_L(m)$ for any $m < m_0$, and no solution passing through $(m_1,\ t_1)$ may lie above $t_U(m)$ for any $m > m_1$, since this would necessitate $\beta > \beta_{\max}$. The two straight lines $t_L(m)$ and $t_U(m)$ are parallel.

This concludes the first stage of the solution. The resulting motion starts from the initial point along the lower limit $t = t_U(m)$ and then abruptly "jumps" at $m = m_1$ from $t_L(m_1)$ to t_1. In other words, maximum thrust $\beta = \beta_{\max}$ is maintained until all the fuel has burnt out, followed by a coasting stage until t_1.

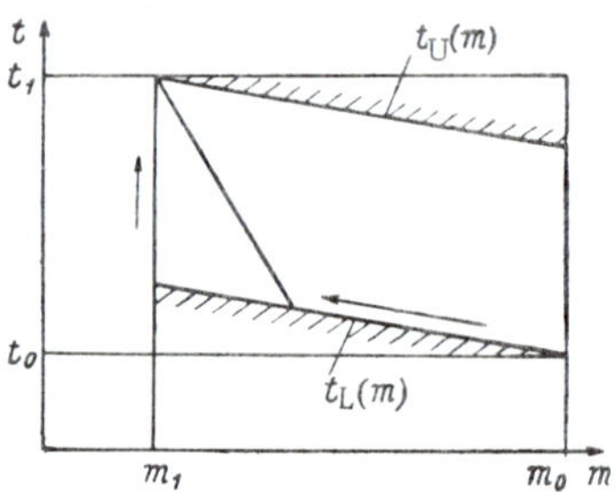

FIGURE 4.1

Here $t(m)$ is a discontinuous function. We can restrain the solution to remain in D by replacing the vertical segment issuing from the point t_1 (see Figure 4.1) with an inclined segment extending to intersection with $t_L(m)$ and making the slope factor of this segment go to infinity $k \rightarrow \infty$. This is achieved by constructing a control sequence with $\beta \rightarrow 0$ over a finite length of time. This sequence satisfies all the conditions of Theorem 1.1.

Indeed, condition (1.12) is satisfied in virtue of (4.14) and condition (1.11) is similarly satisfied, since on this sequence the function R defined by (4.15) is minimum everywhere, except a set of points m whose measure goes to zero.

To find the optimum value of t_1 (the solution of the second stage of the problem), it suffices to integrate (4.1)$-$(4.3) in order to obtain the dependence of h_1 on t_1 and then find the maximum of h_1. It is readily seen

---

* The condition of maximum R (Theorem 1.1) is replaced, as is readily seen, by a condition of minimum when a decreasing argument is substituted for an increasing argument.

that the sought t_1 corresponds to $V_1 = 0$, and the maximum value of h_1 is given by

$$h_1 = \frac{1}{2g} \left(c \ln \frac{m_0}{m_1} \right)^2 - \frac{m_0}{\beta_{max}} c \ln \frac{m_0}{m_1} + \frac{c}{\beta_{max}} (m_0 - m_1). \qquad (4.16)$$

The reader will be able to apply the same method to solve the problem of soft vertical landing with minimum velocity on the surface of a planet without an atmosphere, for given initial conditions.

§ 4. 2. OPTIMUM MOTION CONTROL OF ROCKETS IN
A HOMOGENEOUS GRAVITATIONAL FIELD IN
VACUUM

Various problems of rocket dynamics are concerned with maneuvers which must be performed within a relatively small region of space and in a comparatively short time, so that the real gravitational field acting on the vehicle may be treated as constant (the homogeneous field approximation: the real field depends in general on the position of the spacecraft and the time).

For maneuvers performed near the ground within a sphere of some 100 km radius, the relative error of this approximation is about 1.5%.

As the rocket moves farther from the primary center of attraction, the error decreases (other conditions being constant), so that a midcourse correction on a lunar or an interplanetary trajectory may be treated as taking place in a homogeneous field. This significantly simplifies the analysis and leads to effective solution of numerous optimum problems of rockets dynamics /8, 9/.

The general structure of optimum motion control in a homogeneous field has been investigated by a number of authors /2, 3, 11/.

4.2.1. Statement of the problem

The motion of the center of mass of a rocket powered by a thrust vector in vacuum in a homogeneous gravitational field is described by the following set of equations:

$$r' = -\frac{1}{\beta} V; \qquad (4.17)$$

$$V' = -\frac{c}{m} p - \frac{1}{\beta} g; \qquad (4.18)$$

$$t' = -\frac{1}{\beta} , \qquad (4.19)$$

where $r = (r^1, r^2, r^3)$, $V = (V^1, V^2, V^3)$ are, respectively, the radius vector and the velocity vector of a point in the inertial frame of reference; $g = (g^1, g^2, g^3)$ is the constant gravitational acceleration vector; p is the unit thrust vector (a control element); t is time; m is the mass (independent variable); β is the rate of mass consumption (another control element); c is the nozzle velocity. In addition to $(4.17) - (4.19)$, the admissible motions also satisfy the constraints

$$0 \leqslant \beta \leqslant \beta_{max}, \quad \text{where} \quad \beta_{max} \leqslant \infty \tag{4.20}$$

and the boundary conditions

$$(r_0, \ V_0) \in B(m_0) \tag{4.21}$$

for $t_0 = t_{0f}$ and $m = m_0$, and

$$(r_1, \ V_1) \in B(m_1) \tag{4.22}$$

for $t_1 = t_{1f}$ and $m = m_1$.

Here $B(m)$ is a set in the vector space $(r, \ V)$. The set of control functions

$$z(m) = (r(m), \ V(m), \ t(m), \ p(m), \ \beta(m))$$

satisfying the above conditions will be designated D. Our problem is to find a control function $\bar{z}(m)$ in D on which the functional — some function $F(r_1, \ V_1)$ — attains its minimum.

If the sought control function $\bar{z}(m)$ does not exist in D, we will have to construct a minimizing sequence $\bar{z}_s(m)$ in D.

4.2.2. Transition to Problem 2

We construct the functions

$$R(r, \ V, \ p, \ \beta, \ m) = \varphi_r \left(-\frac{1}{\beta} \right) V + \varphi_V \left(-\frac{c}{m} p - \frac{1}{\beta} g \right) + \varphi_t \left(-\frac{1}{\beta} t \right) + \varphi_m; \tag{4.23}$$

$$\Phi(r_0, \ V_0, \ r_1, \ V_1) = F(r_1, \ V_1) + \varphi(r_1, \ V_1, \ t_1, \ m_1) - \varphi(r_0, \ V_0, \ t_0, \ m_0). \tag{4.24}$$

The first two terms on the left in (4.23) are scalar products of three-dimensional vectors.

The function φ is defined as the general solution of the partial differential equation

$$\varphi_r V + \varphi_V g + \varphi_t = 0, \tag{4.25}$$

where the left-hand side is the coefficient before $\dfrac{1}{\beta}$ in expression (4.23) for the function R. The solution of this equation has the form

$$\varphi = \varphi_1(\xi, \ \eta, \ m), \tag{4.26}$$

where $\xi=(\xi^1,\ \xi^2,\ \xi^3)$, $\eta=(\eta^1,\ \eta^2,\ \eta^3))$ are the vector first integrals of the characteristic system of equation (4.25):

$$\frac{dr}{dt}=V; \tag{4.27}$$

$$\frac{dV}{dt}=g. \tag{4.28}$$

It is readily seen that the characteristic equations in this case describe the coasting of a point in a homogeneous field.

The first integrals ξ, η are given by

$$\xi=V-gt; \tag{4.29}$$

$$\eta=r-Vt+\frac{1}{2}\,gt^2. \tag{4.30}$$

Inserting the solution (4.26) in expressions (4.23), (4.24) for R and Φ and changing over to new variables, we obtain

$$R_1=\min_{\beta} R=-\frac{c}{m}\,\{\varphi_\xi-\varphi_\eta t\}\,p+\varphi_m; \tag{4.31}$$

$$\Phi_1=F_1(\xi_1,\ \eta_1,\ t_1)+\varphi_1(\xi_1,\ \eta_1,\ m_1)-\varphi_1(\xi_0,\ \eta_0,\ m_0). \tag{4.32}$$

The functions R_1 and Φ_1 require further investigation. In accordance with the theory of 3.3.2, this transformation of the functions R and Φ may be interpreted as a transition to Problem 2, which calls for minimization of the same functional over the set D_1 characterized by the same system of boundary conditions and constraints and a new system of differential relations:

$$\xi'=-\frac{c}{m}\,p; \tag{4.33}$$

$$\eta'=\frac{c}{m}\,pt, \tag{4.34}$$

where the components of the vectors ξ, η play the role of phase coordinates, and the components of p and the time t are the control elements. If equations (4.33), (4.34) are supplemented with equation (4.19), the resulting system will be equivalent to equations (4.17) — (4.19). Suitable transformation of coordinates will move one system into the other. The expressions for the functional and the boundary conditions change accordingly under this transformation. Since (4.33), (4.34) contain fewer constraints than (4.17) — (4.19), the set D_1 is wider than D, i.e., $D_1 \supset D$, so that

$$\inf_{D_1} I \leqslant \inf_{D} I. \tag{4.35}$$

We will show in what follows that the equality sign applies in (4.35), i.e., the initial problem may be replaced by Problem 2.

Note that (4.33), (4.34) have a peculiar characteristic: the right-hand sides of these equations are independent of the phase coordinates.

The necessary conditions for a minimum of R are the following:

$$R_{1\xi}(\bar{\xi},\ \bar{\eta},\ \bar{p},\ \bar{t},\ m)=0 \rightarrow \psi'_{(\xi)}=0; \tag{4.36}$$

$$R_{1\eta}(\bar{\xi},\ \bar{\eta},\ \bar{p},\ \bar{t},\ m)=0 \rightarrow \psi'_{(\eta)}=0; \tag{4.37}$$

$$R_1(\bar{\xi},\ \bar{\eta},\ \bar{p},\ \bar{t},\ m)=\inf_{p,\,t\in[t_L,\,t_U]} R(\bar{\xi},\ \bar{\eta},\ p,\ t,\ m), \tag{4.38}$$

where

$$\psi_{(\xi)}=\varphi_\xi(m,\ \bar{\xi},\ \bar{\eta}),$$

$$\psi_{(\eta)}=\varphi_\eta(m,\ \bar{\xi},\ \bar{\eta}).$$

Here $\bar{\xi}(m),\ \bar{\eta}(m)$ are the values of $\bar{\xi},\ \bar{\eta}$ along the sought optimal solution.

Relations (4.36)$-$(4.38) correspond to the adjoint system and the condition of minimum of H (a function with a decreasing argument) in Pontryagin's maximum principle. These equalities thus give the necessary conditions of optimality. By (4.36), (4.37), ψ_ξ, ψ_η are constant (since the right-hand sides of (4.33), (4.34) are independent of the phase coordinates).

The function H has the form

$$H=-\frac{c}{m}\,p\left(\psi_{(\xi)}-\psi_{(\eta)}t\right). \tag{4.39}$$

The minimum of H is evidently attained for

$$p=+\frac{\psi_{(\xi)}-\psi_{(\eta)}t}{|\psi_{(\xi)}-\psi_{(\eta)}\,t|} \tag{4.40}$$

and it is equal to

$$H_1=\min_p H=-\frac{c}{m}\,|\psi_{(\xi)}-\psi_{(\eta)}t|=$$

$$=-\frac{c}{m}\sqrt{(\psi_{(\xi)})^2-2\psi_{(\xi)}\psi_{(\eta)}t+(\psi_{(\eta)})^2 t^2}. \tag{4.41}$$

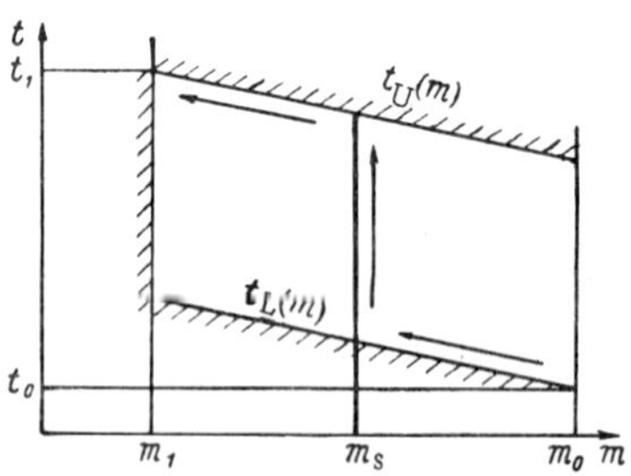

FIGURE 4.2

The minimum of H_1 with respect to t coincides with the maximum of the radicand in (4.41), which is a quadratic trinomial in t. If $\psi_\eta \neq 0$, the coefficient before t^2 is positive, so that the radicand in (4.41) can attain a maximum only for extreme values of t, i. e., $t_L(m)$, $t_U(m)$ (Figure 4.2).

Comparison of the values of H_1 at the lower and the upper limits yield the following conditions for optimum selection with respect to t:

$$t = \begin{cases} t_L(m) & \text{for} \quad \dfrac{\psi_{(\varepsilon)}\psi_{(\eta)}}{(\psi_{(\eta)})^2} > \dfrac{t_L(m) + t_U(m)}{2}, \\[2ex] t_U(m) & \text{for} \quad \dfrac{\psi_{(\varepsilon)}\psi_{(\eta)}}{(\psi_{(\eta)})^2} < \dfrac{t_L(m) + t_U(m)}{2}. \end{cases} \tag{4.42}$$

Since the left-hand side of the inequality in (4.42) is constant and the right-hand side is a monotonically decreasing function of mass, the motion from m_0 to m_1 (in the direction of decreasing mass) can have at most one control change, necessarily from the lower limit $t_L(m)$ to the upper $t_U(m)$.

In particular cases, depending on the values of $\psi_{(\varepsilon)}\dot{\psi}_{(\eta)}$, control functions without abrupt change are possible: either $t = t_L(m)$ or $t = t_U(m)$.

For $\dot{\psi}_{(\eta)} = 0$, the function H_1 is independent of t and the condition of minimum of H_1 with respect to t is satisfied by any piecewise-continuous function $t(m)$ varying between the fixed limits. This case is observed when either the initial or the final value of the vector is free.

If the function F_1 is concave and the boundary conditions are fixed or defined by linear forms in terms of the initial and the final values of the coordinates (the components of the vectors ξ_0, η_0, ξ_1, η_1), the necessary conditions (4.36)−(4.38) are also sufficient, since the right-hand sides of (4.33), (4.34) are independent of the phase coordinates. Therefore each solution satisfying the above conditions is of necessity optimal. In particular, if the limit value η_1 or η_0 of the vector η is free, the optimal solutions are inherently not unique. Any combination of the control functions

$$\bar{p} = \frac{\psi_{(\varepsilon)}}{|\psi_{(\varepsilon)}|}, \quad \bar{t}(m),$$

where $\psi_{(\varepsilon)}$ is selected so that the boundary conditions are satisfied and $\bar{t}(m)$ is any function falling within the admissible limits, is an optimal solution of Problem 2. This can be verified directly, by eliminating equation (4.34) from system (4.33), (4.34) and considering the problem for equation (4.33) only, whose right-hand side is independent of t.

In the general case of nonlinear boundary conditions and nonlinear functionals, conditions (4.36)−(4.38) are only necessary. To find the optimal solution in this case, a further analysis is needed (e. g., further investigation of R). However, since conditions (4.36), (4.38) are necessary, they describe the structure of the optimal solution.

4. 2. 3.   The solution of Problem 2 in the original
set.   The structure of optimal control functions

The solution of Problem 2 belongs to the set $D_1 \supset D$. If this solution in
the (t, m) plane satisfies equation (4.19) with the given constraints on β, it
also belongs to the original set D and is thus a solution of the original
problem. It is readily seen that solutions of the form shown in Figure 4.2
(the case $\psi_{(\eta)} \neq 0$) satisfy equation (4.19). The sections of the upper and
the lower boundaries correspond to $\beta = \beta_{max}$ (motion with maximum thrust),
and the jump at the switching point corresponds to $\beta = 0$ (coasting with the
thrust off). The direction of the thrust vector along the powered sections
is determined by equation (4.40). If we strictly confine the solutions to the
class D, which corresponds to continuous functions $t(m)$, the solution of
the problem is provided by a sequence which in the (t, m) plane (Figure 4.2)
corresponds to a sequence of the functions $t(m)$ obtained when the sudden
jumps at the points m_0, m_1, m_s are replaced by inclined straight sections
with the slope factor going to infinity (as in § 4. 1).

The case $\psi_{(\eta)} = 0$ requires special consideration. The solution of
equations (4.36) − (4.38) in this case incorporates all the piecewise-
continuous functions $t(m)$ satisfying the constraints $t_L(m) \leqslant t(m) \leqslant t_U(m)$ on
$[m_1, m_0]$ and the boundary conditions t_0, t_1. These functions may include those
which do not satisfy equation (4.19). If the conditions are nonlinear at the
two end points, further analysis is required in order to find the optimal
solution of Problem 2. There is danger, however, of ending up with an
optimal solution which corresponds to a function $t(m)$ that does not satisfy
equation (4.19).

In order to assess the imminence of this danger, let us consider the
solution of system (4.33), (4.34) for $\psi_{(\eta)} = 0$. By (4.40) we have

$$p = -\frac{\psi_{(\xi)}}{|\psi_{(\xi)}|} = \text{const.} \tag{4.43}$$

Hence

$$\xi_1 = \xi_0 + p \ln \frac{m_0}{m_1}; \tag{4.44}$$

$$\eta_1 = \eta_0 + p \int_{m_0}^{m_1} \left(-\frac{t(m)}{m} \right) dm. \tag{4.45}$$

We see from (4.44) and (4.45) that the vectors $(\xi_1 - \xi_0)$ and $(\eta_1 - \eta_0)$ are
collinear with the vector p and the ends of the vectors $(\eta_1 - \eta_0)$ define a
segment in the direction p with the end points

$$\left. \begin{aligned} -c \int_{m_0}^{m_1} \frac{t_L(m)}{m} \, dm; \\[2em] -c \int_{m_0}^{m_1} \frac{t_U(m)}{m} \, dm. \end{aligned} \right\} \tag{4.46}$$

The sought optimal value of $(\eta_1 - \eta_0)$ must lie inside this segment. If it coincides with one of the end points, the corresponding function $t_L(m)$ or $t_U(m)$ is unique and is represented by equation (4.19). If, however, it lies inside the interval, there are infinitely many solutions $t(m)$ satisfying the only condition

$$p\,(\eta_1 - \eta_0) = -\,c \int_{m_0}^{m_1} \frac{t\,(m)}{m}\, dm. \qquad (4.47)$$

Among these solutions, we can always choose a set of functions $t(m)$ satisfying equation (4.19) and the appropriate boundary conditions. A suitable function $t(m)$ may be sought in the form

$$t(m) = -\,\frac{1}{\beta_{max}}\, m + a, \qquad (4.48)$$

where the constant a is readily obtained from (4.47). It follows from the above that in any case the solution of Problem 2 incorporates the solution of the original problem, which is readily derived from the former. We thus indeed have the equality sign in (4.35).

4.2.4. Generalization to the case of free end points

The above results significantly depend on the construction of the limit values of t, which can be done fairly easily for the case of fixed end points. The solution of the problem with free end points belongs to the set of solutions of problems with fixed end points and therefore has the same structure as before. The additional conditions of optimum for m_0, t_0, m_1 can be obtained by varying the general solution of Problem 2 with fixed end points with respect to these parameters. The only exceptions are the problems seeking minimum t_1 and maximum m_1: the former since t is a control element in Problem 2, and not a phase coordinate, and the latter since problems with fixed m_1 are meaningless. The corresponding problems can be solved by introducing new variables

$$t = a\tau; \quad m = b\mu,$$

where τ, μ are the new variables (with fixed end points), and a and b are introduced in (4.33), (4.34) as additional phase coordinates with the appropriate supplementary equations $a' = 0$, $b' = 0$. It is readily seen that the structure of the sought solution remains unchanged in this case also.

Indeed, the equations of Problem 2 in this case take the form

$$\left.\begin{aligned}
\xi' &= -\frac{c}{\mu}\, p; \\
\eta' &= \frac{c}{\mu}\, pa\tau; \\
a' &= 0; \\
b' &= 0,
\end{aligned}\right\} \qquad (4.49)$$

where the derivatives are with respect to μ, and the upper and lower limits of τ ($\tau_U(\mu)$ and $\tau_L(\mu)$) are constructed using the equation

$$\tau' = -\frac{b}{a}\frac{1}{\beta}$$

with fixed boundary values τ_0, μ_0, τ_1, μ_1. They are described by the equations

$$\left.\begin{aligned}
\tau_L(\mu) &= \tau_0 - \frac{b}{a}\,\frac{1}{\beta_{\max}}\,(\mu - \mu_0);\\[2mm]
\tau_U(\mu) &= \tau_1 - \frac{b}{a}\,\frac{1}{\beta_{\max}}\,(\mu - \mu_1).
\end{aligned}\right\} \tag{4.50}$$

The functions H and H_1 for this system are obtained by substituting $a\tau$ for t in (4.39) and (4.41). Since the right-hand sides of these equations are independent of ξ and η, the vectors $\psi_{(\xi)}$ and $\psi_{(\eta)}$ are again constant along the optimal solution, so that all the previous conclusions remain in force.

The new system of equations is supplemented with equations for $\psi_{(a)}$ and $\psi_{(b)}$,

$$\psi'_{(a)} = -\frac{\partial H}{\partial a} - \frac{\partial H}{\partial \tau}\,\frac{\partial \tau_{L,U}}{\partial a};$$

$$\psi'_{(b)} = -\frac{\partial H}{\partial \tau}\,\frac{\partial \tau_{L,U}}{\partial b},$$

where the terms with the derivatives $\dfrac{\partial \tau_{L,U}}{\partial a}$ and $\dfrac{\partial \tau_{L,U}}{\partial b}$ account for the dependence of the constraints for τ on a and b. When τ takes other than one of its limit values, these derivatives vanish. For the limit values of τ, the equation of the corresponding boundary is differentiated.

These comments are not to be regarded as practical recommendations for the solution of optimum problems. Their aim is to clarify the general structure of the solution. In practice, these problems may sometimes be solved by considering the conjugate problems, namely the maximization of the coordinate for fixed end points t_0, m_0, m_1, t_1.

4.2.5. Singular control

Sometimes the condition of maximum of H with respect to u does not permit identifying a finite number of potential optimum programs. This situation is encountered, e. g., when one of the control functions enters the Hamiltonian in linear form and its coefficient vanishes in the optimal case. The corresponding cases are known as s i n g u l a r   c o n t r o l (in the sense of the maximum principle). The necessary conditions of the maximum principle for the original problem often do not permit reaching definite conclusions regarding the optimality of these control functions, and this in its turn interferes with the elucidation of the general structure of the optimal control.

Let us consider this aspect in more detail. We will use the equations
of motion in new variables (4.33), (4.34), (4.19), which are equivalent to
(4.17)−(4.19).

The function H for these equations has the form

$$H = -(\psi_{(\varepsilon)} - \psi_{(\eta)}t)\, \frac{c}{m}\, p + \psi_{(t)}\, \frac{1}{\beta}\,,$$

(4.51)

where by the conjugate system $\psi_{(\varepsilon)}$ and $\psi_{(\eta)}$ are constant and $\psi_{(t)}(m)$ is
described by the equation

$$\psi'_{(t)} = -\frac{\partial H}{\partial t} = \psi_{(\eta)}\, \frac{c}{m}\, p.$$

(4.52)

Here $\psi_{(t)}(m)$ is the control switching function.

Along the singular control, p is given by (4.40) and

$$\psi_{(t)} \equiv 0.$$

(4.53)

By (4.52), condition (4.53) leads to

$$(\psi_{(\varepsilon)} - \psi_{(\eta)}t)\, \psi_{(\eta)} = 0.$$

(4.54)

The last equality is satisfied under the following conditions

1) $$t = \mathrm{const} = \frac{\psi_{(\varepsilon)}\psi_{(\eta)}}{\psi_{(\eta)}^{2}}\,;$$

(4.55)

2) $$\psi_{(\eta)} = 0.$$

(4.56)

Condition (4.55) corresponds to $\dfrac{1}{\beta} = 0$ (pulsed control). This is a case
of so-called pulsed singular control. Its optimality can be elucidated by investigating Problem 2 analogous to the original problem, with t treated as
a control element. It is readily seen that (4.55) corresponds to a stationary
point of the quadratic trinomial in t in the radicand in (4.41). Since the
coefficient before t is positive, the Hamiltonian has a maximum (and not
a minimum) with respect to t in Problem 2. The conclusion is that the
singular control of the type being considered is not optimal.

We have seen before that the singular control functions satisfying
condition (4.56) may be optimal and as a rule they are not unique. The
structure of optimal control in this case has been fully elucidated.

4.2.6. Generalization to the case of multistage rockets

The preceding considerations can be generalized to the case of multi-
stage rockets when the engine of each stage has its own maximum thrust.
The argument in this case is conveniently chosen as the characteristic
velocity, which unlike mass is a continuous function of time:

$$u = \int_{0}^{t} \frac{c\beta(t)}{m(t)}\, dt.$$

Equations (4.33), (4.34) take the form

$$\frac{d\xi}{du} = p; \quad \frac{d\eta}{du} = -pt,$$
(4.57)

and equation (4.19) is written as

$$\frac{dt}{du} = \frac{m_i(u)}{c\beta(u)} e^{-\frac{u - u_i(u)}{c}},$$
(4.58)

where $m_i(u)$, $\beta(u)$, $u_i(u)$, $c(u)$ are known piecewise-constant functions of u (u_i, m_i are the characteristic velocity and the mass of the remaining part of the rocket when a stage is jettisoned).

It is readily seen that expression (4.40) for the optimal control function p and the radicand in H_1 do not change in this case. Only the region of admissible values of t changes. It is constructed along the same lines as before; its form is shown in Figure 4.3. Both the upper $t_U(u)$ and the lower $t_L(u)$ limits of t for $u \in (u_0, u_1)$ are solutions of (4.58) for t; $t_L(u)$ passes through the point (t_0, u_0), and $t_U(u)$ through the point (t_1, u_1). Since $t_L(u)$ and $t_U(u)$ are monotonic functions and the interval $(t_L(u), t_U(u))$ is fixed, the previous results remain applicable.

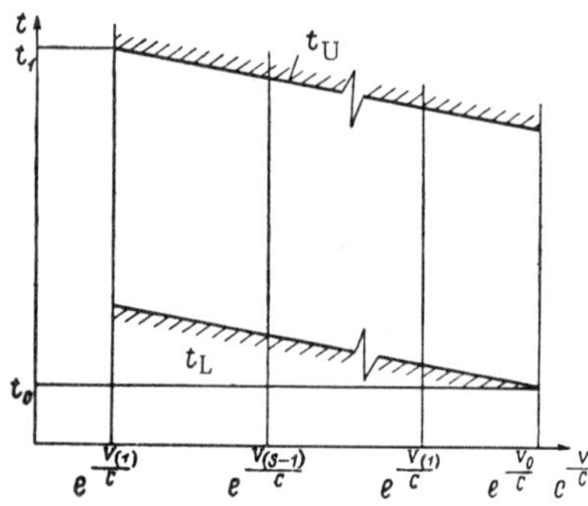

FIGURE 4.3

The above results are generalized to the case when the original equations have the form

$$\left.\begin{aligned}
r' &= -\frac{1}{\beta}(v + f_1(t)); \\[4pt]
v' &= -\frac{c}{m}p - \frac{1}{\beta}f_2(t); \\[4pt]
t' &= -\frac{1}{\beta},
\end{aligned}\right\}$$
(4.59)

where $f_1(t)$, $f_2(t)$ are some piecewise-continuous vector functions such that the integrals

$$\int_{t_0}^{t} f_1(t)\,dt; \quad \int_{t_0}^{t} f_2(t)\,dt$$

exist. In this case, only the expressions of the first integrals ξ, η in terms of r, v, t change. The equations of Problem 2 and the conclusions relating to the realization of its solutions and singular controls remain unchanged.

The general variational problem of the dynamics of a point of varying mass in vacuum in a homogeneous gravitational field thus has been reduced completely to a new Problem 2 for system (4.33), (4.34), with the mass appearing as the argument, the coasting integrals of the point mass as the phase variables, and the thrust direction p and time t as the control elements. The constraints on time are provided by equation (4.19) and the boundary conditions for t and m. The right-hand sides of (4.33), (4.34) are significantly independent of the phase coordinates; as a result, the adjoint factors are constant. The investigation of Problem 2 gave the following results.

1. Optimal control in general may comprise two powered sections of maximum thrust separated by a coasting stage. In the limit, with $\beta_{max} \to \infty$, this corresponds to a case of two-pulse motion, when the first power pulse is applied at the initial instant and the second pulse at the final instant.

2. With special boundary conditions, when the final value of the vector η is either free or is fixed so that the vector increments $\eta_1 - \eta_0$ and $\xi_1 - \xi_0$ are colinear, the optimal solution is not unique. In terms of the maximum principle, these solutions constitute the singular control for the original problem.

3. Pulsed singular control is non-optimal and is not included in the structure of optimal solutions.

4. The structure of optimal solutions described above is fundamentally independent of the number of stages, even if the maximum thrust potential of each stage is different.

5. The equations of optimal control can be readily integrated analytically. Any particular optimization problem for motion in a homogeneous field thus may be solved as a problem of minimum of a function of a finite number of variables, although this is not a general recommendation. In a number of cases, proceeding from the specific features of particular problems, the expressions of the control functions may be markedly simplified before integration, and the final result is thus obtained by a shorter path.

§ 4.3. VERTICAL ASCENT OF A ROCKET IN THE ATMOSPHERE TO MAXIMUM ALTITUDE

The present section is devoted to the solution of a problem presented in Introduction (Examples I.1 and I.2). This problem of rocket dynamics has been treated by a number of authors /4, 10, 13/. On the one hand, it is of immediate practical interest for programming the thrust of sounding rockets, and on the other hand, it may be regarded as a model of more complex, two-dimensional and three-dimensional problems of rocket dynamics. Proceeding from the considerations in the Introduction, we see that solutions satisfying Euler's equation are not necessarily optimal. The analysis that follows not only provides an illustration of the technique but actually solves the problem in full form and proves the optimality of the solutions.

Let us briefly reiterate the problem. Find the motion reaching the maximum altitude among all the admissible motions satisfying the conditions

$$\left.\begin{aligned}
\dot{h}=V; \quad \dot{V}=-\frac{1}{m}(X(h,\,V)-P)-g(h); \\
\dot{m}=-P/c; \\
h(0)=h_0; \quad V(0)=V_0; \quad m(0)=m_0; \quad 0\leqslant P\leqslant\infty,
\end{aligned}\right\} \tag{4.60}$$

where h, V, m are the altitude, the velocity, and the mass of the rocket; $X(h,\,V)$ is the drag; P is the thrust; $g(h)$ is the gravitational acceleration; c is the nozzle velocity.

As is readily seen, the equations of motion (4.60) contain the controlled thrust P in linear form, so that the method of multiple maxima can be applied.

The functions R and Φ for this problem take the form

$$R=\varphi_h V+\varphi_V\left[-\frac{1}{m}X(h,\,V)-g(h)+P/m\right]-\varphi_m P/c+\varphi_t; \tag{4.61}$$

$$\Phi(h_1,\,V_1,\,t_1)=h_1+\varphi(t_1,\,m_1,\,h_1,\,V_1)-\varphi(t_0,\,m_0,\,h_0,\,V_0). \tag{4.62}$$

The equation of multiple maxima in this case reduces to

$$\varphi_V\frac{c}{m}-\varphi_m=0 \tag{4.63}$$

and its general solution is

$$\varphi=\varphi_1(t,\,h,\,\eta(V,\,m)), \tag{4.64}$$

where

$$\eta(V,\,m)=V-c\ln\frac{m_0}{m}. \tag{4.65}$$

The physical meaning of the first integral η is the difference between the actual velocity and the characteristic velocity of the rocket.

The equations of Problem 2 have the form

$$\dot{h}=V; \quad \dot{\eta}=-\frac{1}{m}X(h,\,V)-g(h). \tag{4.66}$$

Seeing that the time of motion is not fixed and η is a monotonic function of time ($\dot{\eta}$ is strictly negative by (4.66)), we may change over to a new variable, dropping the equation for t. The problem thus reduces to investigating the one equation

$$\frac{dh}{d\eta} = -\frac{V}{\dfrac{1}{m}X(h,V)+g(h)} \equiv f(h,\eta,m), \qquad (4.67)$$

where

$$V = \eta + c \ln \frac{m_0}{m}.$$

Now, applying the maximum principle, say, we derive conditions which exhaust the solution of the following Mayer problem: for every fixed η from the interval (η_1, η_0)

$$f\left(\bar{h}(\eta),\ \eta, \overline{m}(\eta)\right) = \inf_{m_{\mathrm{H}} \leqslant m \leqslant m_{\mathrm{B}}} f\left(\bar{h}(\eta),\ \eta,\ m\right) \qquad (4.68)$$

and at the final instant

$$f\left(\bar{h}_1, \eta_1,\ m_1\right) = 0. \qquad (4.69)$$

By (4.69), $V_1 = 0$, i.e.,

$$\eta_1 = -c \ln \frac{m_0}{m_1}.$$

The limits $m_{\mathrm{L}}(\eta)$ and $m_{\mathrm{U}}(\eta)$ are constructed using the equation

$$\frac{dm}{d\eta} = \frac{P/c}{\dfrac{1}{m}X+g(h)}, \qquad (4.70)$$

which is obtained by changing over to the new variable η in the original system (4.60) and in the boundary conditions.

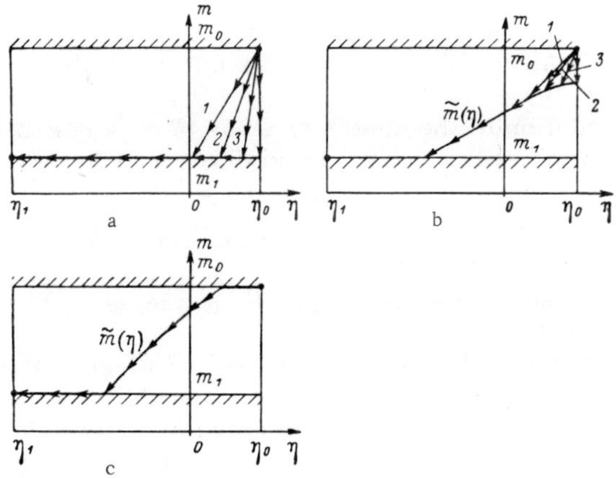

FIGURE 4.4

The constructions in this case are entirely analogous to those considered in § 4.1. The upper and the lower limits $m_U(\eta)$, $m_L(\eta)$ are the solutions of equation (4.70) for $P=0$ which pass through the points (m_0, η_0), (m_1, η_1), respectively. They are shown in Figure 4.4. Evidently, $m_U(\eta_0) \equiv m_0$, $m_L(\eta_1) = m_1$. The solution of equation (4.68) for the general case is shown in Figure 4.4. Depending on particular boundary conditions and rocket parameters, all the three cases shown in Figure 4.4, a, b, c, can be attained. Case a corresponds to a combusion pulse in which all the fuel is burnt, followed by coasting to $\eta_1 (V_1 = 0)$. Case b is a combination of an initial pulse with the fuel burning to some $\tilde{m}(\eta_0)$ then a stage of throttled thrust $\tilde{m}(\eta)$ which satisfies the condition

$$\frac{\partial f}{\partial m} = 0 \tag{4.71}$$

until all fuel has burnt out and coasting to $V=0$. Finally, case c differs from the previous case in one respect only: the boosting pulse is replaced by deceleration down to a velocity $\tilde{V}$, corresponding to the condition

$$\tilde{m}(\eta) = m_0. \tag{4.72}$$

If we are to remain strictly in the set of admissible solutions, the pulsed solutions should be approximated with sequences of control functions with the thrust increasing to infinity (curves 1, 2, 3, ... in Figure 4.4).

Condition (4.71) corresponds to the dependence derived in Examples I.1 and I.2 by differentiating the equation of the control switching line $M=0$. This dependence, in its turn, is equivalent to the relations obtained for throttled thrust programs in /2, 4, 10, 13/.

It is readily seen that the solution of Problem 2 constructed in this way satisfies the original equations and therefore solves the original problem. This solution indicates that the singular control is optimal in the present case.

§ 4.4. THE DEGENERATE PROBLEM OF THE OPTIMAL POWERED ASCENT TRAJECTORY OF AN AIRCRAFT

Consider the motion of the center of mass of an aircraft in the vertical plane under the combined action of the gravitational forces, air drag, and engine thrust. The equations of motion are written in the form /7/

$$m\dot{V} = P(h, V) - X(h, V, \alpha) - mg\sin\theta; \tag{4.73}$$

$$mV\dot{\theta} = P(h, V)\alpha + Y(h, V, \alpha) - mg\cos\theta, \tag{4.74}$$

where h is the altitude, V is the velocity of the aircraft center of mass, m is the mass, θ is the angle of inclination of the trajectory to the horizon, α is the angle of attack, i.e., the angle between the wing chord and the direction of the velocity vector, g is the gravitational acceleration, which is assumed constant in magnitude and direction, $P(h, V)$ is the engine thrust specified as a function of altitude and velocity, X and Y are the drag and the lift on the aircraft, respectively:

$$\left.\begin{aligned} X &= c_x(M,\ \alpha)\ \frac{\varrho(h)\,V^2}{2}\ S; \\[2mm] Y &= c_y(M,\ \alpha)\ \frac{\varrho(h)\,V^2}{2}\ S. \end{aligned}\right\} \qquad (4.75)$$

Here $\varrho(h)$ is the density of the atmosphere, S is the effective wing area, M is the Mach number, defined as the ratio of the velocity V to the speed of sound $a(h)$ at a given altitude, c_x and c_y are the drag coefficients; the dependence of the drag coefficients on the angle of attack for flight in the dense layers of the atmosphere is expressed by the relations

$$c_x(M,\ \alpha) = c_{x0}(M) + b(M)\alpha^2, \qquad (4.76)$$

$$c_y(M,\ \alpha) = c_y^{\alpha}(M)\alpha. \qquad (4.77)$$

Equations (4.73), (4.74) are written assuming sufficiently small angles of attack, and we may therefore take $\cos\alpha \approx 1$ and $\sin\alpha \approx \alpha$. These equations should be supplemented by two further relations

$$\dot{h} = V\sin\theta; \qquad (4.78)$$

$$\dot{m} = -\beta(h,\ V), \qquad (4.79)$$

where β is the per-second fuel consumption (a known function of h and V). We will only consider ascent trajectories with

$$0 \leqslant \theta < \pi. \qquad (4.80)$$

Horizontal flight with $\theta = 0$ is allowed in the limit.

The importance of this conditions for flight near the Earth is quite evident. For many types of aircraft, it is highly desirable over the entire section of powered ascent. Thus, using (4.78) and remembering that because of the inequality $\dot{h} > 0$ we may eliminate the time t from the equations of motion, we change over to the argument h and write

$$\frac{dV}{dh} = \frac{P-X}{mV\sin\theta} - \frac{g}{V}; \qquad (4.81)$$

$$\frac{d\theta}{dh} = \frac{P\alpha + Y}{mV^2\sin\theta} - \frac{g\,\mathrm{ctg}\,\theta}{V^2}; \qquad (4.82)$$

$$\frac{dm}{dh} = -\frac{\beta}{V\sin\theta}. \qquad (4.83)$$

If the aircraft motion is described by these equations, the three numbers V, θ, h may be regarded as a phase vector, and the attack angle α assumes the role of a control element.

The set V_α of the admissible values of the control element α is defined by the inequality

$$|\alpha| \leqslant \gamma, \qquad (4.84)$$

where the number $\gamma>0$ is chosen from considerations of aircraft stability and mechanical strength.

Suppose that at the initial time $t=0$ the aircraft has known altitude, known velocity, and known mass,

$$h(0)=h_0; \quad V(0)=V_0; \quad m(0)=m_0. \tag{4.85}$$

When at a given altitude $h_1>h_0$, the aircraft is required to have a given velocity $V_1>V_0$, i. e.,

$$V(t_1)=V_1. \tag{4.86}$$

These boundary conditions define two sets $V_y(h=h_0)$ and $V_y(h_1)$ of the admissible states of the aircraft at altitudes $h=h_0$ and $h=h_1$. Let us determine the set $V_y(h)$ over the interval $(h_0,\ h_1)$. We assume that the mass of fuel consumed by the entire maneuver is sufficiently small, so that we may take

$$m \approx m_0 = \text{const}. \tag{4.87}$$

The angle θ according to the above is restrained to the limits $0 \leqslant \theta < \pi$. The velocity V should satisfy the inequalities

$$V_1(h) \leqslant V \leqslant V_2(h), \tag{4.88}$$

where the lower limit $V_1(h)$ is defined by the formula

$$V_1(h)=\begin{cases} \widetilde{V}_1(h) & \text{for} \quad \widetilde{V}_1(h) \geqslant 0 \\ 0 & \text{for} \quad \widetilde{V}_1(h) < 0 \end{cases}, \tag{4.89}$$

$\widetilde{V}_1(h)$ is the solution of equation (4.81) for $\alpha=\pm\gamma$, $\theta=\dfrac{\pi}{2}$ with the initial condition $V(h_0)=V_0$. The upper limit is expressed by the equality

$$V_2(h)=\min[\widetilde{V}_2(h),\ V_2^*(h)]. \tag{4.90}$$

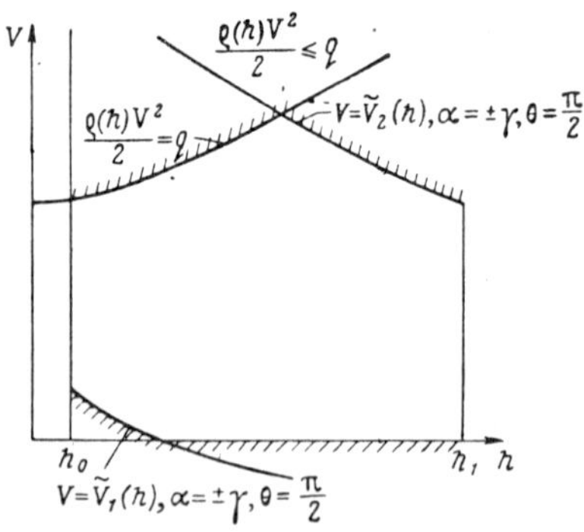

FIGURE 4.5

Here $\bar{V}_2(h)$ is the solution of equation (4.81) for $\alpha=\pm\gamma$, $\theta=\dfrac{\pi}{2}$ which passes through the point h_1, V_1 of the (h, V) plane (Figure 4.5), and $V_2^*(h)$ is obtained from engineering considerations. Thus, considerations of mechanical strength demand that the dynamic head does not exceed some fixed value q:

$$\frac{\varrho(h)V^2}{2} \leqslant q. \tag{4.91}$$

The curves $V=\bar{V}_1(h)$ and $V=\bar{V}_2(h)$ are the projections of the "maximum braking" trajectory onto the (h, V) plane. This "maximum braking" is attained for vertical ascent $\left(\theta=\dfrac{\pi}{2}\right)$ and maximum drag $(\alpha=\pm\gamma)$.

Physically, the condition $\bar{V}_1(h)\leqslant V\leqslant\bar{V}_2(h)$ means that an aircraft moving at an altitude h with velocity $V(h)>\bar{V}_2(h)$, say, will reach the altitude h_1 with velocity $V(h_1)>V_1$, thus violating the boundary condition $V(h_1)=V_1$.

An element z of the set D of admissible trajectories is thus a combination of a piecewise-differentiable vector function $y(h)=(V(h),\ \theta(h))$ and a piecewise-continuous function $\alpha(h)$ defined on the segment $[h_0,\ h_1]$, where they satisfy equations (4.81), (4.82) and conditions

$$\left.\begin{array}{c} y(h)\in V_y(h); \\[4pt] \alpha(h)\in V_\alpha. \end{array}\right\} \tag{4.92}$$

The last conditions are equivalent to the boundary conditions (4.85), (4.86) and inequalities (4.80), (4.84), (4.88).

Among all the elements $z\in D$, find the one which minimizes the functional

$$I(z)=m_0-m(h_1)=\int_{h_0}^{h_1}\frac{\beta}{V\sin\theta}\,dh. \tag{4.93}$$

In other words, we are looking for an admissible flight trajectory such that the ascent from altitude h_0 to h_1 and acceleration from velocity V_0 to V_1 is effected with minimum expenditure of fuel.

Before proceeding with a solution of the problem, we have to make one further assumption concerning the properties of the right-hand sides of equations (4.81), (4.82), which will further restrict the range of application of the problem but significantly simplify its solution. We will assume that within the limits specified by (4.84), the drag coefficient may be treated as independent of the angle of attack:

$$c_x(M,\ \alpha)\equiv c_x^0(M)+b(M)\alpha^2\approx c_x^0(M). \tag{4.94}$$

This assumption makes the problem degenerate (the control element α is now a linear component on the right in the relevant equations), and it can thus be solved by the special methods described in Chapter III.

Thus following the general outline for the solution of variational problems from the sufficient optimum conditions, we introduce a function

$\varphi(h, V, \theta)$ which depends on the argument h and the phase coordinates V and θ. We further construct the functions

$$R(h, V, \theta, a) = \varphi_V \left[\frac{P - X(h, V)}{mV \sin \theta} - \frac{g}{V} \right] +$$

$$+ \varphi_\theta \frac{1}{mV^2 \sin \theta} [P\alpha + Y(h, V, a) - mg \cos \theta] - \frac{\beta(h, V)}{V \sin \theta} + \varphi_h; \qquad (4.95)$$

$$\Phi(V_0, \theta_0, V_1, \theta_1) = \varphi(h_1, V_1, \theta_1) - \varphi(h_0, V_0, \theta_0). \qquad (4.96)$$

We are looking for the vector functions $(\overline{V}(h), \overline{\theta}(h), \overline{a}(h)) \in D$ and a function $\varphi(h, V, \theta)$ such that

$$R[h, \overline{V}(h), \overline{\theta}(h), \overline{a}(h)] = \mu(h) \equiv \sup_{\substack{|a| \leq \gamma \\ V_1(h) \leq V \leq V_2(h) \\ 0 < \theta < \pi}} R(h, V, \theta, a); \qquad (4.97)$$

$$\Phi(V_0, \overline{\theta}_0, V_1, \overline{\theta}_1) = \inf_{\substack{0 < \theta_0 < \pi \\ 0 < \theta_1 < \pi}} \Phi(V_0, V_1, \theta_0, \theta_1). \qquad (4.98)$$

In virtue of (4.75), (4,77), and (4.94), R is linear in a. We may thus use the method of multiple maxima in order to find $\varphi(h, V, \theta)$. If R is to be independent of a, we should have

$$\varphi_\theta(h, V, \theta) = 0. \qquad (4.99)$$

The general solution of this equation is

$$\varphi = \varphi_1(h, V), \qquad (4.100)$$

where $\varphi_1(h, V)$ is an arbitrary function of h and V.

Inserting $\varphi = \varphi_1(h, V)$ in (4.95) and (4.96), we find

$$R(h, V, \theta) = \left[\varphi_{1V} \frac{P - X}{m} - \beta \right] \frac{1}{V \sin \theta} - \varphi_{1V} \frac{g}{V} + \varphi_{1h}; \qquad (4.101)$$

$$\Phi(V_0, \theta_0, V_1, \theta_1) = \varphi_1(h_1, V_1) - \varphi(h_0, V_0). \qquad (4.102)$$

Condition (4.98) is now satisfied automatically for any $\varphi_1(h, V)$, since Φ is independent of θ_0 and θ_1, whereas V_0 and V_1 are fixed.

The function R is now indeed independent of a, i. e., any a maximizes this function for fixed h, V, and θ. We see from (4.101) that by imposing an additional condition on $\varphi = \varphi_1(h, V)$,

$$\varphi_{1V} \frac{P - X}{m} - \beta = 0, \qquad (4.103)$$

we make R independent of θ either. Condition (4.103) imposed on $\varphi_1(h, V)$ corresponds to a repeated application of the method of multiple maxima to our problem, since θ in Problem 2 is a control element.

From (4.103) we have

$$\varphi_{1V} = \frac{m\beta\,(h,\,V)}{P\,(h,\,V) - X\,(h,\,V)}\,; \qquad (4.104)$$

$$\varphi_1(h,\,V) = \int_{V_0}^{V} \frac{m\beta}{P-X}\,dV\,; \qquad (4.105)$$

$$\varphi_{1h}(h,\,V) = m\int_{V_0}^{V} \frac{\partial}{\partial h}\left[\frac{\beta}{P-X}\right]dV\,. \qquad (4.106)$$

The integral in (4.105) and (4.106) is evaluated for fixed h from an arbitrary lower limit V_0. Inserting (4.104), (4.105), (4.106) in (4.101), we find

$$R\,(h,\,V) = -\frac{mg\beta\,(h,\,V)}{(P-X)\,V} + m\int_{V_0}^{V} \frac{\partial}{\partial h}\left(\frac{\beta}{P-X}\right)dV\,. \qquad (4.107)$$

The right-hand side of this equality is a know function of h and V. Condition (4.97) for a maximum of R with the given φ defines the optimal dependence $\overline{V}(h)$:

$$R(h,\,\overline{V}(h)) = \sup_{V_1(h)\,\leqslant V\,\leqslant V_2(h)} R(h,\,V). \qquad (4.108)$$

Here $R(h,\,V)$ is given by (4.107).

It follows from (4.108) that the optimal flight program $\overline{V}(h)$ for every fixed h either coincides with one of the limit values $V_1(h)$ or $V_2(h)$, or is a solution of the equation

$$R_V\,(h,\,V) \equiv -\frac{\partial}{\partial V}\left[\frac{mg\beta}{(P-X)\,V}\right] + m\frac{\partial}{\partial h}\left[\frac{\beta}{P-X}\right] = 0, \qquad (4.109)$$

or, finally, corresponds to a vertical jump (in the ($h,\,V$) plane) from one of these $\overline{V}(h-0)$ to another $\overline{V}(h+0)$ corresponding to horizontal flight. The order in which these pieces are "matched" is determined by equation (4.108), which fully describes the optimal program.

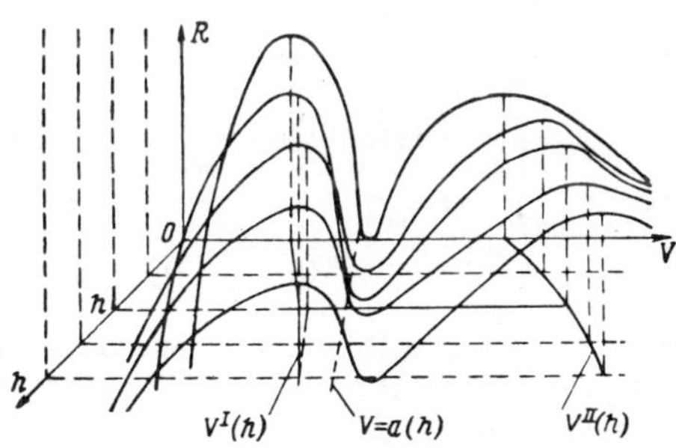

FIGURE 4.6

Let us now consider the engineering computations of the optimal powered ascent trajectory proceeding from (4.108).

1. Numerical integration gives the solutions $\hat{V}_1(h)$ and $\hat{V}_2(h)$ of equation (4.81) for the vertical ascent $\left(\theta=\frac{\pi}{2}\right)$, which pass through the points (h_0, V_0), and (h_1, V_1), respectively. These solutions are plotted in the (h, V) plane. The curve of constant dynamic head q is plotted in the same plane,

$$\frac{\varrho(h) V^2}{2} = q,$$

and using (4.88), (4.89) we construct the limits $V_1(h)$ and $V_2(h)$ of the admissible velocities.

2. For every fixed $h \in [h_0, h_1]$, we use (4.107) to construct the function $R(h, V)$ on the segment $[V_1(h), V_2(h)]$. The integral in (4.107) is obtained by numerical or graphical integration. Figure 4.6 shows a specimen function of this kind.

3. For every h, we choose a velocity $\bar{V}(h)$ corresponding to the maximum value of $R(h, V)$ over $[V_1(h), V_2(h)]$. As a rule (see Figure 4.6), the function $R(h, V)$ has two local maxima satisfying equation (4.109) (the subsonic and the supersonic extremals, respectively); for $h<h^*$ (low altitudes), $R[h, V^{\mathrm{I}}(h)]>R[h, V^{\mathrm{II}}(h)]$, and for $h>h^*$ (high altitudes), $R[h, V^{\mathrm{I}}(h)]<R[h, V^{\mathrm{II}}(h)]$. There may, however, be more than two maxima.

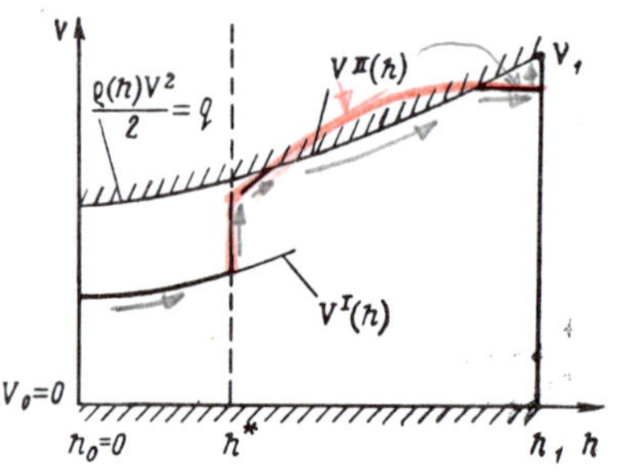

FIGURE 4.7

The optimal dependence $\bar{V}(h)$ corresponding to the function $R(h, V)$ obtained in this case is shown in Figure 4.7 for the initial conditions $h_0=0$, $V_0=0$. For $h<h^*$,

$$\bar{V}(h) = V^{\mathrm{I}}(h),$$

and for $h>h^*$,

$$\bar{V}(h) = \min[V^{\mathrm{II}}(h), V_2(h)].$$

This $\bar{V}(h)$ corresponds to the following flight conditions. The aircraft first accelerates along a horizontal trajectory from $V_0=0$ to some velocity $V^{\mathrm{I}}(0)$ at the same altitude $h=0$ (in practice, obviously, the flight altitude

is somewhat higher). The aircraft then climbs at the subsonic extremal
velocity $V = V^{\mathrm{I}}(h)$ to the altitude $h = h^*$, where $V^{\mathrm{I}}(h^*) \neq V^{\mathrm{II}}(h^*)$. At this
altitude, the aircraft again assumes a horizontal trajectory, accelerating
from $V^{\mathrm{I}}(h^*)$ to $V^{\mathrm{II}}(h^*)$, covering the entire sonic range of velocities. Then
the aircraft climbs along the supersonic extremal $V^{\mathrm{II}}(h)$ until the limit of
the dynamic head constraint is reached. After that it moves with constant
dynamic head q, again along the supersonic extremal, and finally
accelerating in the horizontal direction at altitude h_1 to velocity V_1.

A different optimal trajectory $\overline{V}(h)$ is shown in Figure 4.8. Sections of
this trajectory may correspond, in the limit, to vertical ascent of the aircraft.

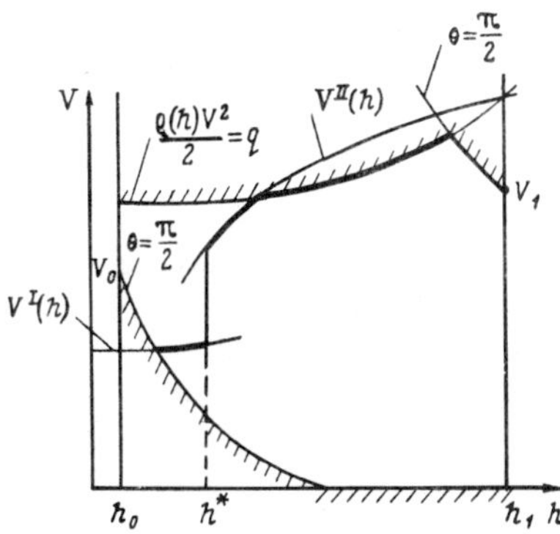

FIGURE 4.8

Inserting the optimal dependence $\overline{V}(h)$ in (4.81), we find the corresponding optimal dependence $\overline{\theta}(h)$.

The kinks in $\overline{V}(h)$ correspond to discontinuities in $\overline{\theta}(h)$. Strictly speaking,
this solution, although exact in the sense of Problem 2, does not exist in the
original class D, since the functions $\theta(h)$ in this class are continuous.

A solution which belongs to D can be obtained if the above solution of
Problem 2 is replaced in the neighborhood of the discontinuity points by
solutions of the original system of differential equations with $\alpha = +\gamma$ for the
discontinuities where $\overline{\theta}$ increases and with $\alpha = -\gamma$ for the discontinuities
where $\overline{\theta}$ decreases.

The solution formed in this way should be treated as an approximate
solution in D, with the estimate $|I_2 - I^*|$, where I_2 is the minimum value of
the functional in Problem 2, I^* is the value of the functional for the new
solution. Since $I_2 \leqslant \overline{I}$, where $\overline{I}$ is the exact minimum of the functional on D,
we have

$$|I_2 - I^*| \geqslant |\overline{I} - I^*|.$$

If the left-hand side of this inequality is sufficiently small, the right-hand
side is also small, i. e., the approximate solution is sufficiently exact and no
better approximation need be sought. If the estimate $|I_2 - I^*|$ is large and
does not satisfy the required accuracy, the above method cannot be applied in
undiluted form to the solution of the particular problem, and it only provides a

qualitative description of the sought solution. In these cases, the result may be used, say, as the first approximation in some successive approximations scheme whereby the solution is brought to an end.

R e m a r k 1. A similar procedure can be applied to find a trajectory $\overline{V}(h)$ minimizing the maneuver time

$$T = \int_{h_0}^{h_1} \frac{1}{V \sin \theta} \, dh. \qquad (4.110)$$

To this end, $\beta(h, V)$ should be replaced with unity in expression (4.109) for the function $R(h, V)$.

R e m a r k 2. The degenerate problem of the optimum powered ascent of an aircraft considered in this section played an important role in the development of optimum flight theory. It was first tackled by Ostoslavskii and Lebedev /6/ in 1946. They obtained the subsonic extremal $V^{\mathrm{I}}(h)$, known as Ostoslavskii's curve. Miele /12/, by an ingenious application of Green's theorem, proved the optimality of trajectories containing one of the extremal branches and described the fundamental structure of the optimal trajectory in the case of multiple extremals. The question of the optimal position of the "jumps" (the horizontal flight sections) between the different extremals remained open, however. A complete solution of the problem by the method described in this section was finally published in /5/.

§ 4.5. APPROXIMATE SYNTHESIS OF OPTIMAL ANGLE-OF-ATTACK CONTROL FOR A LIQUID-PROPELLANT WINGED AIRCRAFT

4.5.1. Statement of the problem. Equations of motion

The object of our analysis in this section is a liquid-propellant winged aircraft normally flying in the dense layers of the atmosphere, at altitudes up to 20 km. The motion of the aircraft is investigated in the wind system of coordinates under the following basic assumptions.

1. The aircraft motion is considered in a vertical plane in the atmosphere of a non-rotating Earth with a homogeneous gravitational field.

2. The aircraft is treated as a point of varying mass, since liquid-propellant engines are characterized by high rates of fuel consumption.

3. Since the engine thrust is assumed to remain constant, the aircraft motion is entirely determined by controlling the angle of attack of the wings.

4. The per-second consumption of the liquid propellant under conditions of constant engine thrust is independent of flight velocity and flight altitude /6/, i.e., the aircraft mass is a known function of time.

5. The engine thrust is constant, and the direction of the thrust vector is along the longitudinal axis of the aircraft.

6. The polar curve of the aircraft is a family of quadratic parabolas of the form

$$c_x = c_{x0}(M) + c(M)\alpha^2; \\
c_y = A(M)\alpha, \qquad\qquad (4.111)$$

where c_x is the drag coefficient, c_{x0} is the component of the drag coefficient for zero angle of attack, c is a coefficient which allows for the lift-drag coupling, α is the angle of attack, M is the Mach number.

The functions $c_{x0} = f(M)$; $c = f(M)$; $A = f(M)$ are represented as polynomials in M.

7. The aerodynamic forces on the aircraft are determined assuming stationary conditions.

8. The limits of variation of the angle of attack are sufficiently small and we may therefore take

$$\sin\alpha \approx \alpha; \quad \cos\alpha \approx 1 - \frac{\alpha^2}{2}.$$

The equations of motion of an aircraft in the wind system of coordinates under the above assumptions take the form

$$\dot{V} = \frac{1}{m}\,[P - X_0(V,h) - (0.5P + X_i(V,h))\alpha^2 - mg\sin\theta]; \qquad (4.112)$$

$$\dot{\theta} = \frac{1}{Vm}\,\{[P + Y(V,h)]\,\alpha - mg\cos\theta\}; \qquad (4.113)$$

$$\dot{h} = V\sin\theta, \qquad (4.114)$$

where V is the flight velocity in km/sec, h is the flight altitude in km, θ is the angle of inclination of the flight trajectory to the horizon in rad, P is the engine thrust, $X_0(V, h)$ is the drag for zero angle of attack, $X_i(V, h)\alpha^2$ and $Y(h, V)\alpha$ are, respectively, the induced drag and the lift.

The atmospheric density and the speed of sound as a function of altitude are described by the following relations:

$$\varrho = 0.125\,\frac{20 - h}{20 + h}; \quad a(h) = 0.340\,\sqrt{1 - 0.02255h}\,.$$

$$\varrho = 0.2105e^{-\frac{h}{6340}}; \quad a(h) = 0.295 \quad (\text{for } h > 11\,\text{km}).$$

In what follows α is considered as a control element, with the following inequality constraints:

$$\alpha_1 \leqslant \alpha \leqslant \alpha_2. \qquad (4.115)$$

We will construct an optimal synthesis of the angle of attack $\alpha(t, y) = \bar{\alpha}(t, h, V, \theta)$ which moves any point of the space (t, h, V, θ) into a point t_{1f}, h_{1f}, θ_{1f}, V_1 with the maximum velocity V_1.

Here $t_{1f} = 100\,\text{sec}$, $h_{1f} = 19,000\,\text{m}$, $\theta_{1f} = 0.174\,\text{rad}$.

This is clearly the dual problem of the problem of minimum time (or minimum fuel consumption, since the dependence $m(t)$ for reaching the point t_1, h_{1f}, θ_{1f}, V_{1f} is known).

Instead of a problem with a fixed right end point, we will consider a problem with a free right end point for the functional

$$I = V_1 + \sum_{i=2}^{3} \lambda_i (y_1^i - y_{1f}^i)^2, \qquad (4.116)$$

where $\lambda_{2,3}$ are some positive constants.

4. 5. 2. Changing over to new coordinates. Description of $V_y(t)$

The set $V_y(t)$ is defined as a parallelepiped in the y space, described by the following inequalities:

$$\left. \begin{aligned} &\overline{V}(t) \leqslant V \quad : \\ &\overline{\theta} - \varepsilon \leqslant \theta \leqslant \overline{\theta} + \varepsilon; \\ &\overline{h} - \varkappa \leqslant h \leqslant \overline{h} + \varkappa, \end{aligned} \right\} \qquad (4.116a)$$

where

$$\overline{V} = 0.000066t^2 + 0.0027t; \qquad (4.117)$$

$$\overline{\theta} = 0.0002163t^2 - 0.03875t + 1.92; \qquad (4.118)$$

$$\overline{h} = 0.0013t^2 + 0.04824t; \qquad (4.119)$$

$$V = 0.2 \text{ km/sec}; \; \varepsilon = 0.1 \text{ rad}; \; \varkappa = 1 \text{ km}.$$

This definition of $V_y(t)$ is dictated by tentative estimates of that region of the (t, y) space where the optimal trajectories corresponding to the sought synthesis lie (we do not intend to consider this problem in any detail here).

Given this region $V_y(t)$, we can change over to new coordinates (t, z) using the transformations

$$y^i = \overline{y}^i(t) + z^i, \qquad (4.120)$$

where $\overline{y}^i(t)$ are defined by $(4.117) - (4.119)$.

The equations of motion thus take the form

$$\dot{z}^1 \equiv f^1(t, z, \alpha) = \frac{1}{m(t)} \{ P - X_0(\overline{V} + z^1, \overline{h} + z^3) - $$

$$- [0.5P + X_i(\overline{V} + z^1, \overline{h} + z^3)] \alpha^2 - mg \sin(\overline{\theta} + z^2) \} - 0.000132t - 0.0027; \quad (4.121)$$

$$\dot{z}^2 \equiv f^2(t, z, \alpha) = \frac{1}{(\overline{V} + z^1) m(t)} \{ [P + Y(\overline{V} + z^1, \overline{h} + z^3)] \alpha - $$

$$- mg \cos(\overline{\theta} + z^2) \} - 0.0004326t + 0.03875; \qquad (4.122)$$

$$\dot{z}^3 \equiv f^3(t, z, \alpha) = (\overline{V} + z^1) \sin(\overline{\theta} + z^2) - 0.0026t - 0.04824. \qquad (4.123)$$

4.5.3. Solution of the problem

By applying the method of Chapter III, we reduce the solution of the problem to finding a function $\varphi(t, z)$ which satisfies the following conditions of an absolute maximum of the functional:*

$$P(t,z)=\inf_{a\in[a_1,a_2]} R(t,z,a)=\inf_{a\in[a_1,a_2]} [\varphi_z f(t,z,a)+\varphi(t)]=c(t); \qquad (4.124)$$

$$\Phi(t_1,z_1)=z^1+\Sigma\lambda_i\,(z_1^i+z_{13}^i)^2+\varphi(t_1,z_1)=\text{const}, \qquad (4.125)$$

in other words, we have to solve the Cauchy problem for a first-order non-linear partial differential equation.

No exact solution of this problem can be obtained in general. We will therefore use the approximate method of optimal synthesis described in § 2.3. The function φ is sought in the form of a polynomial

$$\varphi(t,z)=\sum_{i_1=0}^{2}(z^1)^{l_1}\left(\sum_{i_2=0}^{2}(z^2)^{l_2}\left(\sum_{i_3=0}^{2}\psi_{i_1i_2i_3}(t)(z^3)^{l_3}\right)\right) \qquad (4.126)$$

with unknown coefficients $\psi_{l_1l_2l_3}(t)$. There is a total of 27 coefficients here. We further demand that conditions (4.124), (4.125) be satisfied for all t at 27 specially constructed reference points in $V_y(t)$ (and definitely not everywhere in this region), which are obtained by combining the following values of the variables z^i:

$$\begin{array}{llll}
z^1: & 0, & \gamma, & 2\gamma; \\
z^2: & -\varepsilon, & 0, & +\varepsilon; \\
z^3: & -\varkappa, & 0, & +\varkappa.
\end{array} \qquad (4.127)$$

In this case the reference curves — the loci of reference points — are parallel straight lines $z_\beta=\text{const}$, $\beta=1, 2, \ldots, 27$, in the (t, z) space.

The function R may be represented in the following form, which is particularly convenient for minimization with respect to a:

$$R=Aa^2+Ba+C+\varphi_t, \qquad (4.128)$$

where

$$A=-\varphi_{z^1}\left[0.5\,\frac{P}{m}+X_i(t,\overline{V}+z^1,\overline{h}+z^3)\right]; \qquad (4.129)$$

$$B=\varphi_{z^2}\left[\frac{1}{\overline{V}+z^1}\left(\frac{P}{m}+Y(t,\overline{V}+z^1,\overline{h}+z^3)\right)\right]; \qquad (4.130)$$

$$C=\varphi_{z^1}\left[\frac{P}{m}-X_0(t,\overline{V}+z^1,\overline{h}+z^3)-g\sin(\overline{\theta}+z^2)-\right.$$

$$\left.-0.000132t-0.0027\right]-\varphi_{z^2}\left[\frac{g\cos(\overline{\theta}+z^2)}{\overline{V}+z^1}+\right.$$

$$+0.000432t-0.03875\Big]+\varphi_{z^3}[(\overline{V}+z^1)\sin(\overline{\theta}+z^2)-0.0026t-0.04824]; \quad (4.131)$$

φ_{z^i} and φ_t are partial derivatives of the polynomial (4.126) which, after expansion of the parentheses and transition to a continuous numeration of the unknowns, take the form

---

_* In maximizing the functional, the condition of maximum of R in Theorem 1.1 is replaced by a condition of minimum.

$$\varphi = \psi_1 + \psi_2 z^1 + \psi_3 z^2 + \psi_4 z^3 + \psi_5 (z^1)^2 + \psi_6 (z^2)^2 + \psi_7 (z^3)^2 +$$
$$+ \psi_8 z^1 z^2 + \psi_9 z^1 z^3 + \psi_{10} z^2 z^3 + \psi_{11} z^1 (z^2)^2 + \psi_{12} z^1 (z^3)^2 +$$
$$+ \psi_{13} z^2 (z^3)^2 + \psi_{14} (z^1)^2 z^2 + \psi_{15} (z^1)^2 z^3 + \psi_{16} (z^2)^2 z^3 +$$
$$+ \psi_{17} (z^1)^2 (z^2)^2 + \psi_{18} (z^1)^2 (z^3)^2 + \psi_{19} (z^2)^2 (z^3)^2 + \psi_{20} z^1 z^2 z^3 +$$
$$+ \psi_{21} z^1 z^2 (z^3)^2 + \psi_{22} z^1 (z^2)^2 z^3 + \psi_{23} (z^1)^2 z^2 z^3 +$$
$$+ \psi_{24} z^1 (z^2)^2 (z^3)^2 + \psi_{25} (z^1) z^2 (z^3)^2 + \psi_{26} (z^1)^2 (z^2)^2 z^3 + \psi_{27} (z^1)^2 (z^2)^2 (z^3)^2. \qquad (4.132)$$

Minimization of the quadratic trinomial (4.128) in α under the above constraints on α gives the following results, which depend on the value of the trinomial coefficients:

$$P(t, z) = \inf_{\alpha} R(t, z, \alpha) = \mathscr{H}(t, z) + \varphi(t), \qquad (4.133)$$

where for $A \geqslant 0$,

$$\left. \begin{array}{lll} \mathscr{H} = A\alpha_1^2 + B\alpha_1 + C & \tilde{\alpha} = \alpha_1 & -\dfrac{B}{2A} \leqslant \alpha_1; \\[2ex] \mathscr{H} = -\dfrac{B^2}{A} + C & \tilde{\alpha} = -\dfrac{B}{2A} & \alpha_2 < -\dfrac{B}{2A} > \alpha_1; \\[2ex] \mathscr{H} = A\alpha_2^2 + B\alpha_2 C & \tilde{\alpha} = \alpha_2 & -\dfrac{B}{2A} > \alpha_2; \end{array} \right\} \qquad (4.134)$$

for $A < 0$,

$$\left. \begin{array}{ll} \mathscr{H} = A\alpha_1^2 + B\alpha_1 + C & \tilde{\alpha} = \alpha_1; \\[1ex] \mathscr{H} = A\alpha_2^2 + B\alpha_2 + C & \tilde{\alpha} = \alpha_2, \end{array} \right\} \qquad (4.135)$$

and the coefficients A, B, C are obtained from $(4.129) - (4.131)$.

Since (4.124) must be satisfied on the reference lines z_β, we obtain the relations

$$\varphi_{t\beta} = -\mathscr{H}_\beta \ (\beta = 1, 2, \ldots, 27), \qquad (4.136)$$

which constitute a system of first-order nonlinear ordinary differential equations. These equations are not solved for the derivatives $\dot{\psi}_\beta$, but these derivatives enter the equations in linear form. Using matrix notation, we may write (4.136) in the form

$$A\dot{\psi}_\beta = -\mathscr{H}_\beta, \qquad (4.137)$$

where A is the matrix of the system coefficients, $\mathscr{H}_\beta$ is the column vector of the free terms, $\dot{\psi}_\beta$ is the column vector whose elements are the unknowns.

The solution of the above system of equations with the reference lines $z_\beta = \text{const}$ is obtained without considerable difficulty, since the elements of the direct coefficient matrix A are constant at any time $t \in (t_0, t_1)$. To obtain the solution, we first solve the system for $\dot{\psi}$,

$$\dot{\psi} = -A^{-1}\mathscr{H}, \qquad (4.138)$$

and since the elements of the direct coefficient matrix A are constant at any time, the inverse matrix A^{-1} should be determined only at the very beginning when solving (4.137).

The initial conditions for ψ_β ($\beta = 1, 2, \ldots, 27$) at time $t = t_1$ are obtained from (4.125):

$$\Phi(t_1, z_1^i) = z_1^1 + \lambda_2(\widetilde{z^3} - z_1^3)^2 + \lambda_1(\widetilde{z^2} - z_1^2)^2 + \varphi(t_1, z_1^1, z_1^2, z_1^3) = C, \qquad (4.139)$$

or

$$\Phi(t_1, z_1^1, z_1^2, z_1^3) = z_1^1 + \lambda_1[(\widetilde{z_1^2})^2 - 2\widetilde{z_1^2}z_1^2 + (z_1^2)^2] + \lambda_2[(\widetilde{z_1^3})^2 -$$
$$- 2\widetilde{z^3}z_1^3 + (z_1^3)^2] + \psi_1 + \psi_2 z_1^1 + \psi_3 z_1^2 + \psi_4 z_1^3 +$$
$$+ \psi_5(z_1^1)^2 + \psi_6(z_1^2)^2 + \psi_7(z_1^3)^2 + \psi_8 z_1^1 z_1^2 +$$
$$+ \psi_9 z_1^1 z_1^3 + \psi_{10} z_1^2 z_1^3 + \psi_{11} z_1^1 (z_1^2)^2 + \psi_{12} z_1^1 (z_1^3)^2 +$$
$$+ \psi_{13} z_1^2 (z_1^3)^2 + \psi_{14}(z_1^1)^2 z_1^2 + \psi_{15}(z_1^1)^2 z_1^3 +$$
$$+ \psi_{16}(z_1^2)^2 z_1^3 + \psi_{17}(z_1^1)^2 (z_1^2)^2 + \psi_{18}(z_1^1)^2 (z_1^3)^2 +$$
$$+ \psi_{19}(z_1^2)^2 (z_1^3)^2 + \psi_{20} z_1^1 z_1^2 z_1^3 + \psi_{21} z_1^1 z_1^2 (z_1^3)^2 +$$
$$+ \psi_{22} z_1^1 (z_1^2)^2 z_1^3 + \psi_{23}(z_1^1)^2 z_1^2 z_1^3 + \psi_{24} z_1^1 (z_1^2)^2 (z_1^3)^2 +$$
$$+ \psi_{25}(z_1^1)^2 z_1^2 (z_1^3)^2 + \psi_{26}(z_1^1)^2 (z_1^2)^2 z_1^3 + \psi_{27}(z_1^1)^2 (z_1^2)^2 (z_1^3)^2 = 0. \qquad (4.140)$$

Condition (4.125) in the above case may be exactly satisfied if we request that

$$\Phi_{z_1^1} = 1 + \psi_2 + 2\psi_5 z_1^1 + \psi_8 z_1^2 + \psi_9 z_1^3 + \psi_{11}(z_1^2)^2 + \psi_{12}(z_1^3)^2 +$$
$$+ 2\psi_{14} z_1^1 z_1^2 + 2\psi_{15} z_1^1 z_1^3 + 2\psi_{17} z_1^1 (z_1^2)^2 + 2\psi_{18} z_1^1 (z_1^3)^2 + \psi_{20} z_1^2 z_1^3 +$$
$$+ \psi_{21} z_1^2 (z_1^3)^2 + \psi_{22}(z_1^2)^2 z_1^3 + 2\psi_{23} z_1^1 z_1^2 z_1^3 + \psi_{24}(z_1^2)^2 (z_1^3)^2 +$$
$$+ 2\psi_{25} z_1^1 z_1^2 (z_1^3)^2 + 2\psi_{26} z_1^1 (z_1^2)^2 z_1^3 + 2\psi_{27} z_1^1 (z_1^2)^2 (z_1^3)^2 = 0; \qquad (4.141)$$

$$\Phi_{z_1^2} = 2\lambda_1 z_1^2 + \psi_3 - 2\lambda_1 \widetilde{z^2} + 2\psi_6 z_1^2 + \psi_8 z_1^1 + \psi_{10} z_1^3 + 2\psi_{11} z_1^1 z_1^2 +$$
$$+ \psi_{13}(z_1^3)^2 + \psi_{14}(z_1^1)^2 + 2\psi_{16} z_1^2 z_1^3 + 2\psi_{17}(z_1^1)^2 z_1^2 + 2\psi_{19} z_1^2 (z_1^3)^2 +$$
$$+ \psi_{20} z_1^1 z_1^3 + \psi_{21} z_1^1 (z_1^3)^2 + 2\psi_{22} z_1^1 z_1^2 z_1^3 + \psi_{23}(z_1^1)^2 z_1^3 +$$
$$+ 2\psi_{24} z_1^1 z_1^2 (z_1^3)^2 + \psi_{25}(z_1^1)^2 (z_1^3)^2 + 2\psi_{26}(z_1^1)^2 z_1^2 z_1^3 +$$
$$+ 2\psi_{27}(z_1^1)^2 z_1^2 (z_1^3)^2 = 0; \qquad (4.142)$$

$$\Phi_{z_1^3} = 2\lambda_2 z_1^3 - 2\lambda_2 \widetilde{z^3} + \psi_4 + 2\psi_7 z_1^3 + \psi_9 z_1^1 + \psi_{10} z_1^2 +$$
$$+ 2\psi_{12} z_1^1 z_1^3 + 2\psi_{13} z_1^2 z_1^3 + \psi_{15}(z_1^1)^2 + \psi_{16}(z_1^2)^2 + 2\psi_{18}(z_1^1)^2 z_1^3 +$$
$$+ 2\psi_{19}(z_1^2)^2 z_1^3 + \psi_{20} z_1^1 z_1^2 + 2\psi_{21} z_1^1 z_1^2 z_1^3 + \psi_{22} z_1^1 (z_1^2)^2 +$$
$$+ \psi_{23}(z_1^1)^2 z_1^2 + 2\psi_{24} z_1^1 (z_1^2)^2 z_1^3 + 2\psi_{25}(z_1^1)^2 z_1^2 z_1^3 + \psi_{26}(z_1^1)^2 (z_1^2)^2 +$$
$$+ 2\psi_{27}(z_1^1)^2 (z_1^2)^2 z_1^3 = 0. \qquad (4.143)$$

It follows from $(4.141)-(4.143)$ that at time $t=t_1$ the functions $\psi_\beta\,(t_1)$ should take on the values

$$\psi_1 = -\lambda_1(\widetilde{z_1^2})^2 - \lambda_2(\widetilde{z_1^3})^2, \quad \psi_2 = 1, \quad \psi_3 = 2\lambda_1\widetilde{z}_1^2,$$

$$\psi_4 = 2\lambda_2\widetilde{z}_1^3, \quad \psi_6 = -\lambda_1, \quad \psi_7 = -\lambda_2, \tag{4.144}$$

$$\psi_5 = \psi_8 = \psi_9 = \ldots = \psi_{27} = 0.$$

Here λ_1 and λ_2 are some positive numbers, which are sufficiently large in our sense.

The solution of system (4.136) with the above boundary conditions gives the unknown coefficients $\psi_\beta(t)$ of the polynomial (4.126) and thus determines the sought approximate expression of the function φ, its derivatives, and finally (from (4.134) and (4.135)) the expression for $\widetilde{a}(t,\ z)$.

Simultaneously with the integration of (4.136), we can integrate the equation

$$\dot{\Delta} = \sup_{V_y} R(t,\ z) - \inf_{V_y} P(t,\ z) \tag{4.145}$$

with the initial condition $\Delta(t_1)=0$. This gives an estimate Δ of the approximate synthesis at the initial time.

4. 5. 4. A scheme for a computer

The problem was solved on the BESM-2 computer. The special program used in this case consists of two parts: the direct program, which solves the system of ordinary differential equations (4.136) and determines the error in the direction from $t_1 = 100$ sec to $t_0 = 30$ sec, and the inverse program, which is used to construct the optimal trajectory for a certain characteristic combination of initial conditions.

The error of the method, the optimal trajectory constructed, and the corresponding optimal program for the angle of attack $\widetilde{a}(t)$ are shown in Figures 4.9 through 4.13. The following initial conditions were use: $t_0 = 30$ sec, $V(t_0) = 0.139\,\mathrm{km/sec}$, $\theta(t_0) = 1.115\,\mathrm{rad}$, $h = 2.7\,\mathrm{km}\,(z^1 = z^2 = z^3 = 0)$.

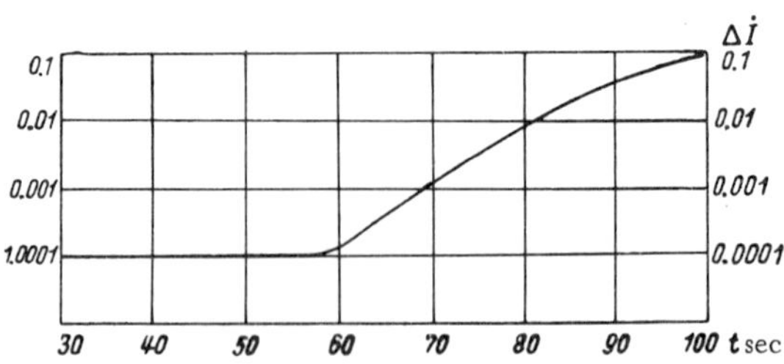

FIGURE 4.9

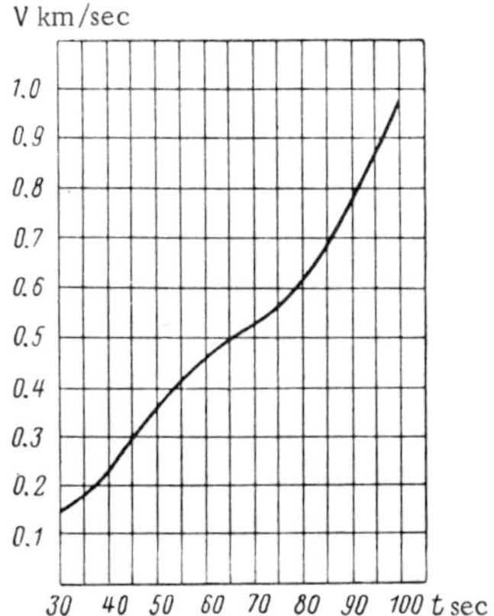

FIGURE 4.10

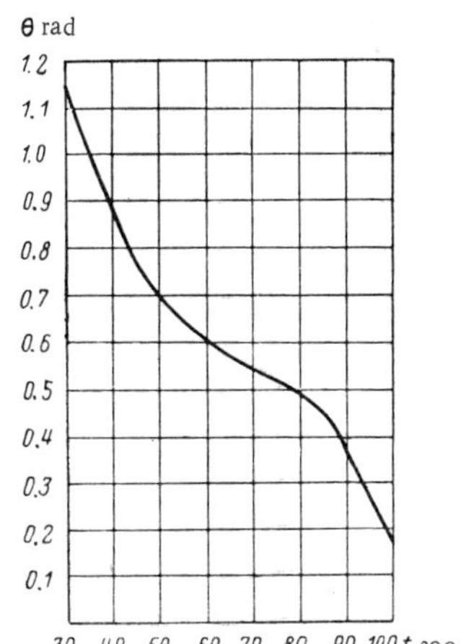

FIGURE 4.11

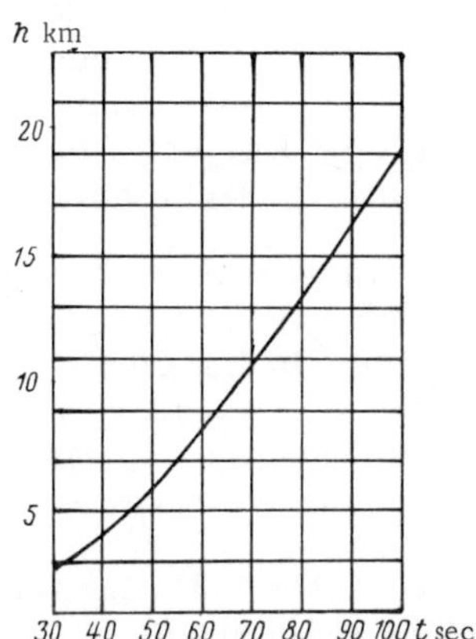

FIGURE 4.12

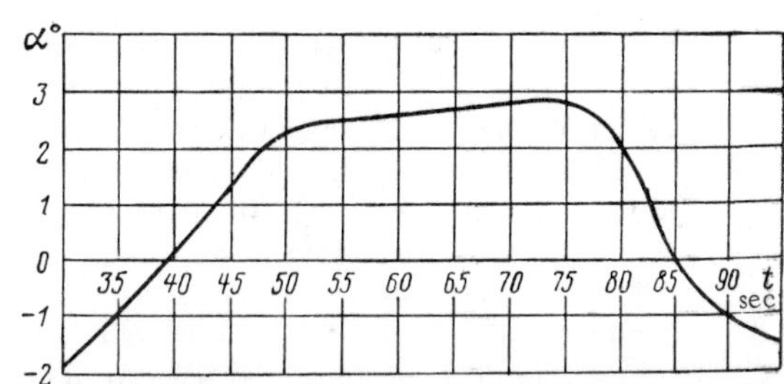

FIGURE 4.13

Bibliography for Chapter IV

1. Bukreev, V.Z. and V.Ya. Moskalenko. Priblizhennyi sintez optimal'nogo upravleniya v zadachakh dinamiki poleta (Approximate Synthesis of Optimal Control in Flight Dynamics Problems).— In: "Issledovaniya po dinamike poleta." Mashinostroenie. 1969.

2. Gurman, V.I. Struktura optimal'nykh rezhimov dvizheniya raket v odnorodnom gravitatsionnom pole (Structure of Optimal Policies of Rocket Motion in a Homogeneous Gravitational Field). — Kosmicheskie Issledovaniya, No. 4, No. 6. 1966.

3. Isaev, V.K. Printsip maksimuma L. S. Pontryagina i optimal'noe programmirovanie tyagi raket (Pontryagin's Maximum Principle and Optimal Rocket Thrust Control). — Avtomatika i Telemekhanika, 22, No. 8. 1961.

4. Kosmodem'yanskii, A.A. Mekhanika tel peremennoi massy (Mechanics of Bodies of Varying Mass). — Izdatel'stvo VVIA im. N.E. Zhukovskii. 1946.

5. Krotov, V.F. and V.N. Sargin. Ob optimal'nykh traektoriyakh pod"ema samoleta (Optimal Aircraft Ascent Paths). — In: "Voprosy analiticheskoi i prikladnoi mekhaniki," V.V. Dobronravov, ed. Oborongiz. 1968.

6. Ostoslavskii, I.V. and A.A. Lebedev. O raschete pod'ema skorostonogo samoleta (Computing the Ascent of a High-Speed Aircraft). — Tekhnika Vozdushnogo Flota, Nos. 8 — 9. 1946.

7. Ostoslavskii, I.V. and I.V. Strazheva. Dinamika poleta. Traektorii letatel'nykh apparatov (Flight Dynamics. Aircraft Trajectories). — Oborongiz. 1963.

8. Okhotsimskii, D.E. K teorii dvizheniya raket (Theory of Rocket Motion). — Prikladnaya Matematika i Mekhanika, 10, No. 2. 1946.

9. Okhotsimskii, D.E. and T.M. Eneev. Nekotorye variatsionnye zadachi, svyazannye s zapuskom iskusstvennogo sputnika Zemli (Some Variational Problems Associations with the Launching of Artificial Earth Satellites). — Uspekhi Fizicheskikh Nauk, No. 1a. 1957.

10. Goddard, R.A. Method of Reaching Extreme Altitudes. — Smithsonian Inst. Publ., Misc. Coll., 71, No. 2. 1919.

11. Leitmann, G. On a Class of Variational Problems in Rocket Flight. — J. Aero/Space Sci., 26, No. 9. 1959.

12. Miele, A. Problemi di minito tempo nel volo non-stazionario degli aeroplani. — Atti. accad. sci. Torino, classe sci. fis.-mat. e nat., 85. 1950 — 1957.

13. Tsien, H.S. and R.C. Evans. Optimum Thrust Programming for a Sounding Rocket. — ARS Journal, 21, No. 5. 1951.

Chapter V

OPTIMAL PROBLEMS IN THE DYNAMICS
OF COASTING IN THE ATMOSPHERE

The present chapter deals with the general variational problem of the dynamics of two-dimensional (plane) motion in the atmosphere of a winged aircraft with controllable aerodynamic lift; constraints on the coordinates and the control element are taken into consideration. We will analyze a typical case when the sought optimal control is degenerate (the case of sliding control). It is shown that if the limit values of the control function are equal in magnitude, the analysis of the sliding control reduces to minimizing an integral functional without differential constraints. The results of this analysis are then applied to solve the problem of the optimal descent of an aircraft in the atmosphere with minimum added integral heat in the critical zone of the aircraft. The method of multiple maxima described in Chapter III is used.

§ 5.1. OPTIMAL TWO-DIMENSIONAL COASTING
OF WINGED AIRCRAFT IN THE ATMOS-
SPHERE IGNORING RANGE

5.1.1. Statement of the problem

We will consider the motion of a winged aircraft, with its engines cut off, in the atmosphere of a spherical non-rotating Earth under the following conditions:

1) equations of motion of the center of mass of the aircraft:

$$\dot{h} = V \sin \theta; \tag{5.1}$$

$$\dot{V} = -X(h, V, c_y) - g \sin \theta; \tag{5.2}$$

$$\dot{\theta} = \frac{1}{V} \left[Y(h, V, c_y) + \left(\frac{V^2}{r_E + h} - g \right) \cos \theta \right]; \tag{5.3}$$

2) boundary conditions

$$h(0),\ V(0),\ \theta(0)) \in B(0);\ (h(t_1),\ V(t_1),\ \theta(t_1)) \in B(t_1); \tag{5.4}$$

3) constraints, for $t \in (0, t_1)$,

$$(h, V, \theta) \in B(t), \quad c_y \in [c_{y_1}(t, h, V), \ c_{y_2}(t, h, V)]. \tag{5.5}$$

Here h, V, θ are respectively the altitude, the velocity, and the inclination of the aircraft trajectory (the phase coordinates); c_y is the lift coefficient of the aircraft (a control element); $Y(h, V, c_y)$, $X(h, V, c_y)$ are the lift and the drag per unit mass of the aircraft; g is the gravitational acceleration; r_E is the Earth's radius.

Equations $(5.1)-(5.3)$ are considered ignoring the horizontal flight range. The functions $X(h, V, c_y)$, $Y(h, V, c_y)$ are assumed to be described by the ordinary equations of stationary aerodynamics:

$$X(h, V, c_y) = \frac{Sg}{G} \frac{\varrho V^2}{2} (c_{x_0}(h, V) + b(h, V, c_y));$$

$$Y(h, V, c_y) = \frac{Sg}{G} \frac{\varrho V^2}{2} c_y,$$

where ϱ is the density of the atmosphere, S is the characteristic area, G is the aircraft weight; for every fixed h and V, $b(h, V, c_y)$ is a symmetrical concave function of c_y. The variation of the gravitational acceleration g with altitude is ignored. The last assumption, like the neglect of the Earth's nonsphericity and its spin, introduce very slight errors in the computation results, and yet greatly simplify the mathematics. These effects can always be restored in full without any modification of the basic mathematical apparatus. The change in aircraft mass associated with possible heat subtraction is also ignored. The functions $h(t), V(t), \theta(t)$ are continuous and piecewise-differentiable, $c_y(t)$ is a bounded piecewise-continuous function. These assumptions are commonly used in variational problems of flight dynamics.

Let D be the set of the admissible motion programs $h(t)$, $V(t)$, $\theta(t)$, $c_y(t)$ satisfying the various conditions and assumptions.

Our problem can be stated as follows. Find a sequence of motion programs

$$\{\bar{h}_s(t), \ \bar{V}_s(t), \ \bar{\theta}_s(t), \ \bar{c}_y(t)\} \in D \tag{5.6}$$

on which the functional

$$I = \int_0^{t_1} f^0(h, V) \, dt + F(h_0, V_0, \theta_0, h_1, V_1, \theta_1) \tag{5.7}$$

goes to its least value over the set D.

The form of the functional (5.7), the constraints (5.5), and the boundary conditions (5.4) depend on the particular problem being considered. They will be improved and adjusted when necessary. Without loss of generality, we may take $c_{y1} < c_{y2}$.

5.1.2. Analysis of the optimum conditions

We will now investigate the function R for the given system comprising the functional and the constraints,

$$R = \varphi_h V \sin\theta - \varphi_V (X(h, V, c_y) + g\sin\theta) +$$

$$+ \varphi_\theta \frac{1}{V}\left[Y(h, V, c_y) + \left(\frac{V^2}{r_E} - g\right)\cos\theta\right] - f^0(h, V) + \varphi_t. \tag{5.8}$$

If the function $\varphi(t, h, V, \theta)$ is given, the function R for some fixed t, h, V, θ will depend on c_y only, and its character, in virtue of the properties of the functions $X(h, V, c_y)$ and $Y(h, V, c_y)$, will be determined primarily by the magnitude and the signs of the coefficients of φ_V and φ_θ. Suppose that the following conditions are observed in some region of the (t, h, V, θ) space:

$$\varphi_V \leqslant 0; \tag{5.9}$$

$$\varphi_V [X(h, V, c_{y_1}) - X(h, V, c_{y_2})] - \varphi_\theta \frac{1}{V}[Y(h, V, c_{y_1}) - Y(h, V, c_{y_2})] = 0. \tag{5.10}$$

As is readily seen, R attains a maximum for two values of c_y, namely $c_{y1}(t, h, V)$ and $c_{y2}(t, h, V)$. If the sought optimal program belongs to the relevant region and satisfies conditions (5.9), (5.10), it constitutes degenerate (sliding) control with degeneracy k of at least 1. If conditions (5.9), (5.10) are not satisfied, the maximum of R is attained only on one c_y, either inside the segment $[c_{y1}, c_{y2}]$ or at one of the end points c_{y1}, c_{y2}. The corresponding optimum control is of Euler type in the former case and of boundary type in the latter. We only investigate the conditions corresponding to degenerate programs since, first, as we shall see from what follows, they are quite typical of the problems being considered, and, second, Euler and boundary optimal controls have been widely treated in the literature $/2 - 5/$.

Following the scheme described in Chapter III (the method of multiple maxima), we will find the general solution of the partial differential equation (5.10). To this end, we first determine the independent first integral of the system

$$\frac{dV}{d\tau} = -[X(h, V, c_{y_1}) - X(h, V, c_{y_2})]; \tag{5.11}$$

$$\frac{d\theta}{d\tau} = \frac{1}{V}[Y(h, V, c_{y_1}) - Y(h, V, c_{y_2})], \tag{5.12}$$

which has the form

$$\xi = \theta + \int K(t, h, V)\, dV, \tag{5.13}$$

where

$$K(t, h, V) = \frac{1}{V}\frac{Y_1 - Y_2}{X_1 - X_2}; \tag{5.14}$$

$$X_{1,2} = X(h, V, c_{y1,2}(h, V, t)); \tag{5.15}$$

$$Y_{1,2} = Y(h, V, c_{y1,2}(h, V, t)). \tag{5.16}$$

The general solution of equation (5.10) is an arbitrary continuous and differentiable function

$$\varphi = \varphi_1(t, h, \xi(t, \theta, h, V)). \tag{5.17}$$

Inserting this function in expression (5.8) for R, we find

$$R_1 = \varphi_{1h} V \sin\theta - \varphi_{1\xi} \{K(t, h, V)[X(h, V, c_y) + g\sin\theta] +$$
$$+ \frac{1}{V}\left[Y(h, V, c_{y_1}) + \left(\frac{V^2}{r_E} - g\right)\cos\theta\right]\} - f^0(h, V) + \varphi_{1t}, \tag{5.18}$$

or

$$R_1 = \varphi_{1h} V \sin\theta - \varphi_{1\xi} \frac{1}{V}\left\{\frac{X_1Y_2 - X_2Y_1}{X_2 - X_1} +\right.$$
$$\left. + K(t, h, V)g\sin\theta - \left(\frac{V^2}{r_E} - g\right)\cos\theta\right\} - f^0(h, V) + \varphi_{1t}, \tag{5.19}$$

where V and θ are related by (5.13)

Here h, ξ are the phase coordinates (the arguments of the function φ_1) and V and θ are the control elements (they are related by (5.13)). Using (5.13), we can eliminate θ. The function R_1 for every fixed t will then depend on h, ξ and V, with V acting as a control element.

Further investigation of the function R_1 in general form does not achieve any useful purpose. One of the ways for further solution is to change over to Problem 2 and minimize the functional (5.7) with the following differential constraints:

$$\dot{h} = V \sin\theta; \tag{5.20}$$

$$\dot{\xi} = \frac{1}{V} \frac{X_1Y_2 - X_2Y_1}{X_2 - X_1} - K(t, h, V)g\sin\theta + \left(\frac{V^2}{r_E} - g\right)\cos\theta, \tag{5.21}$$

where

$$\theta = \xi - \int K(t, h, V)dV.$$

As we could have expected, this problem is simpler than the original problem since the order of system (5.20), (5.21) is 1 less than the order of (5.1) — (5.3).

Problem 2 can be solved by the method of approximate optimal synthesis (see § 2.3). This method can be applied directly to the original problem (if the constraints on the phase coordinates are sufficiently simple), but it involves integration of a system of differential equations of order $m + n$ where n is the order of the system of constraints, and m is the degree of the approximating polynomial. Therefore, even if the order n is lowered by 1 only, the result is a substantial reduction of the overall computational work.

Once Problem 2 has been solved, we have to check the applicability of conditions (5.4), (5.5) and establish that the solution indeed belongs to the angle $\omega(t)$. The latter is verified by inserting the solution $\tilde{h}(t)$, $\tilde{V}(t)$, $\tilde{\theta}(t)$ in the equation

$$\dot{\theta}=\frac{1}{V}\left[vY_1+(1\mp v)Y_2+\left(\frac{V^2}{r_E}-g\right)\cos\theta\right] \tag{5.22}$$

which is then solved for v. If all the above conditions are satisfied, the solution of Problem 2 defines the zero closeness function of the sliding control. The original problem is thus solved. Otherwise, as we have seen before, the function φ is constructed by a different method. Although in this case the solution of Problem 2 does not solve Problem 1, it nevertheless provides some information concerning the type of the sought solution. Using this information, we can construct a reference solution which is subsequently improved by some scheme of successive approximations, e. g., by the method of approximate optimal synthesis for the original problem. A clever choice of the reference trajectory reduces the region in which the optimal synthesis is constructed and thus lowers the degree of the approximating polynomial. As before, this leads to a marked reduction in the volume of computations.

5.1.3. A special case of a symmetric constraint on the lift coefficient

Let

$$|c_{y1}(t, h, V)|=|c_{y2}(t, h, V)|=c_{y\lim}(t, h, V). \tag{5.23}$$

In this important case, the analysis of our general problem can be continued much further. Since the function $X(h, V, c_y)$ is symmetrical, we have

$$X(h, V, c_{y1})=X(h, V, c_{y2}),$$

and condition (5.10) takes the form $\varphi_\theta=0$. This means that the function φ is independent of θ. The function R_1 takes the form

$$R_1=\varphi_{1h}V\sin\theta-\varphi_{1V}[X(h, V, c_{y\lim}(t, h, V))+g\sin\theta]-f^0(h, V)+\varphi_{1t}. \tag{5.24}$$

It follows from the structure of φ_1 and R_1 that $\sin\theta$ is the control element in this case: it enters the expression for R_1 in linear form. Problem 2, in its turn, is degenerate and it can be solved by the method of multiple maxima. We define $\varphi_1(t, h, V)$ so that the coefficient before $\sin\theta$ in (5.24) identically vanishes,

$$\varphi_{1h}V-\varphi_{1V}g=0. \tag{5.25}$$

The general solution of the partial differential equation (5.25) is an arbitrary continuous and differentiable function

$$\varphi_1 = \varphi_2(t,\ \eta(h, V)), \qquad (5.26)$$

where

$$\eta = h + \frac{V^2}{2g} \qquad (5.27)$$

is the independent first integral of the system

$$\left.\begin{aligned}
\frac{dh}{d\tau} &= V;\\[2mm]
\frac{dV}{d\tau} &= -g.
\end{aligned}\right\} \qquad (5.28)$$

Inserting this solution in the expression for R_1, we find

$$R_2 = -\varphi_{2\eta} X(h, V, c_{y\,\lim}(t, h, V)) - f^0(h, V) + \varphi_{2t}, \qquad (5.29)$$

where h, V, and η satisfy (5.27). In (5.29), η is the phase coordinate, and h and V are the control elements (they are related to η by (5.27)). We can change over to one independent control element, h say, by expressing V from (5.27) in terms of η and h:

$$V = \sqrt{2g(\eta - h)}.$$

As in the general case, it is better at this stage to pass to a modified problem (we call it Problem 3) minimizing the functional (5.7) with a single differential constraint

$$\dot{\eta} = -\frac{V}{g}\, X(h, V, c_{y\,\lim}(t, h, V)), \qquad (5.30)$$

where h, V, θ satisfy (5.27). Problem 3 can be solved by approximate optimal synthesis, which in the present case virtually reduces to integrating a system of first-order ordinary differential equations where the number of equations is equal to the degree of the approximating polynomial plus 1 (one-dimensional synthesis).

The resulting solution $(\hat{h}(t),\ \hat{V}(t))$ of Problem 3 should be inserted in the equations of Problem 2 in order to obtain $\sin\theta$. If the result satisfies the constraints, $(\hat{h}(t),\ \hat{V}(t),\ \hat{\theta}(t))$ is a solution of Problem 2. We treat this solution as before.

In practice, the constraints on c_y and the phase coordinates in problems of this kind are generally determined by physical constraints on overloading, temperature, etc., which depend on the flight altitude and velocity and are independent of time. In numerous problems the final time t_1 is not given (e.g., the problem of the minimum added heat in the critical zone, which is considered below). In this case, the right-hand side of equation (5.30) is time-independent. Using this fact and further remembering that any physically admissible motion of the aircraft with the engine cut off, described by equations (5.1)—(5.3), is constrained by the inequality $\dot{\eta} \leqslant 0$ (the total energy of the aircraft cannot increase with time), we may change over from t to a new independent variable η, without distorting the physics of the problem. Problem 3 is thus finally reduced to the problem of minimizing the functional

$$I = - \int_{\eta_0}^{\eta_1} \frac{f^0(h, V)}{\dfrac{V}{g} X(h, V, c_y)} \, d\eta + F(h_0, V_0, \theta_0, h_1, V_1, \theta_1), \qquad (5.31)$$

where

$$V = \sqrt{2g(\eta - h)},$$

without differential constraints and with inequality constraints of the form (5.5) imposed on the free functions $h(\eta)$, $c_y(\eta)$ and on the parameters θ_0 and θ_1; in other words, the general problem is reduced to an elementary problem of the absolute minimum of a function of a single variable for every fixed η. Let $f_1^0(\eta, h, c_y)$ denote the integrand in (5.31). The solution of Problem 3 then coincides with the solution of the equations

$$f_1^0\left(\eta, \tilde{h}(\eta), \tilde{c}_y(h, \eta)\right) = \inf_{\substack{h \in B(\eta) \\ |c_y| < c_{y\lim}(h, \eta)}} f_1^0(\eta, h, c_y), \quad \eta \in (\eta_1, \eta_0); \qquad (5.32)$$

$$F\left(\tilde{h}, V_0(\tilde{h}_0, \eta_0), \tilde{\theta}_0, \tilde{h}_1, V_1(\tilde{h}_1, \eta_1), \tilde{\theta}_1\right) =$$
$$= \inf_{\substack{(h_0, \theta_0) \in B_0 \\ (h_1, \theta_1) \in B_1}} (h_0, V_0(h_0, \eta_0), \theta_0, h, V(h_1, \eta_1), \theta_1), \qquad (5.33)$$

where $B(\eta)$, B_0, B_1 are the regions of the admissible values of h, θ_0, θ_1 defined by conditions (5.5), or the regions of those h, θ_0, θ_1 which are actually attainable for system (5.1)—(5.3) under conditions (5.4), (5.5).

Although the solution of Problem 3 is elementary, we have to consider its existence in the set D. The structure of the functional in Problem 3 is so simple that it permits drawing some important conclusions concerning the feasibility of the solution of this problem. Indeed, let us find the maximum of the integrand in (5.31) with respect to c_y for any admissible fixed η and h. Note that the corresponding optimal value $c_y^*(\eta, h)$ depends only on the sign of the function $f_1^0(h, \eta, c_y)$. If $f_1^0 < 0$, there is only one optimal value of c_y, $c_y^* = 0$. If $f_1^0 > 0$, the minimum of f_1^0 is attained for two different values of c_y,

$$c_y^* = \pm c_{y\lim}(h, \eta).$$

In the negative case, the substitution $c_y = c_y^*$ in the original system determines, in combination with the initial conditions, the unique solution of system (5.1)—(5.3), i. e., a ballistic trajectory which does not necessarily coincide with the solution $(\tilde{h}(\eta), \tilde{V}(\eta), \tilde{\theta}(\eta))$ obtained under completely different conditions.

129

In the positive case, the solution of Problem 3 does not necessarily satisfy equations $(5.1)-(5.3)$ for $c_y=c_y^*$. It only has to satisfy the constraints on $\sin \theta$ and to lie within the angle ω constructed at each point of the trajectory. In this case the solution can be realized by a sliding program — a succession of control functions, each comprising alternating flight sections with maximum and minimum angle of attack.

The denominator of the integrand in (5.31) for any values of the arguments is definite positive, so that the sign of the fraction f_1^0 coincides with the sign of the numerator, i. e., it is determined entirely by the particular form of the function $f^0(h, V)$ (the given optimality criterion). Thus, if $f^0(h, V)\equiv 1$ (the problem of minimum descent time), the solution of Problem 3 is indeed feasible.

If, on the other hand, $f^0(h, V)\equiv-1$ (the problem of maximum lifetime), the solution of Problem 3, strictly speaking, does not exist in D and an alternative method has to be used.

If $f^0(h, V)$ changes its sign, the solution of Problem 3 is strictly feasible in that region of the (h, V) space where $f^0(h, V)>0$.

5.1.4. Generalization to the case of additional control elements

The preceding results are generalized without much difficulty to the case when the angle of attack c_y is not the only control element: other common control elements are the angle of bank γ and the drag coefficient c_{x0}. In practice, c_{x0} can be controlled with the aid of braking apparatus in the form of skirts, flaps or air brakes, and parachutes.

Let us consider the previous problem of minimizing the functional (5.7) over the set D of flight programs satisfying the following conditions.

1. The equations of motion

$$\dot{h}=V \sin \theta; \tag{5.34}$$

$$\dot{V}=-X(h, V, c_{x_0}, c_y)-g \sin \theta; \tag{5.35}$$

$$\dot{\theta}=\frac{1}{V}\left[Y(h, V, c_y)\cos \gamma+\left(\frac{V^2}{r_E}-g\right)\cos \theta\right], \tag{5.36}$$

where h, V, θ are the phase coordinates; c_y, γ, c_{x0} are the control elements.

2. Boundary conditions (5.5), with free end time t_1.

3. Constraints, for $t \in (0, t_1)$,

$$\left.\begin{array}{l}(h, V, \theta)\in B; \\ (c_{x_0}, c_y)\in Q(h, V).\end{array}\right\} \tag{5.37}$$

The angle of bank γ is either not constrained ($-1\leqslant\cos \gamma\leqslant+1$), i. e., equations $(5.34)-(5.36)$ are in fact the projections of the three-dimensional equations of motion on the vertical plane, or takes on the values 0 and π ($\cos \gamma=\pm1$) only, i. e., the aircraft is truly constrained to perform two-dimensional (or almost two-dimensional) motion.

We will accept all the assumptions listed in 5.1.1.

Note that owing to the presence of the additional control element γ, we may treat the product $Y(h,\ V,\ c_y)\cos\gamma$ as an independent linear control element Y_a which assumes values inside the interval

$$[-Y_{a_{\lim}},\ +Y_{a_{\lim}}],$$

where $Y_{a_{\lim}}$ may depend on h and V.

The function R takes the form

$$R=\varphi_h V \sin\theta-\varphi_V\left[X(h,\ V,\ c_{x_o},c_y)+g\sin\theta\right]+\varphi_\theta\ \frac{1}{V}\left[Y_a+\left(\frac{V^2}{r_E}-g\right)\cos\theta\right]+\varphi_t. \tag{5.38}$$

Successively applying the method of multiple maxima and changing over to a new argument, we end up, as before, with Problem 3 minimizing the functional

$$I_1=-\int_{\eta_0}^{\eta_1}\frac{f^0(h,V)}{\dfrac{V}{g}\,X(h,V,c_{x_o},c_y)}\,d\eta+F(h_0,\ V_0,\ \theta_0,\ h_1,\ V_1,\ \theta_1), \tag{5.39}$$

where

$$V=\sqrt{2g(\eta-h)}.$$

For every $\eta\in(\eta_1,\ \eta_0)$, the solution $\tilde{h}(\eta),\ \tilde{c}_{x0}(\eta),\ \tilde{c}_y(\eta)$ of Problem 3 is defined by the condition

$$f_1^0(\eta,\ \tilde{h}(\eta),\ \tilde{c}_{x_o}(\eta),\ \tilde{c}_y(\eta))=\inf_{\substack{h\in B_1(\eta)\\(c_{x0},\ c_y)\in Q_1(h,\ \eta)}} f_1^0(\eta,\ h,\ c_{x_o},\ c_y), \tag{5.40}$$

where $f_1^0(\eta,\ h,\ c_{x0},\ c_y)$ is the integrand in (5.39). In other words, for every $\eta\in(\eta_1,\eta_0)$, the values of $\tilde{h},\tilde{c}_{x0},\tilde{c}_y$ are obtained as the point of absolute minimum of the function $f_1^0(\eta,\ h,\ c_{x0},\ c_y)$ with respect to all the three independent variables $h,\ c_{x0},\ c_y$.

Owing to the additional control element γ, the transition to Problem 3 does not require any assumptions concerning equal limits of variation of the angle of attack. In general, the solution of Problem 3 exists in D when $\tilde{c}_y(\eta)$ is unique, provided that $\tilde{c}_y(\eta)\neq0$.

Indeed, inserting $\tilde{h},\ \tilde{c}_y,\ \tilde{c}_{x0}$ in the initial conditions, we find $\tilde{\theta}(\eta)$ and then some function $Y_a(\eta)$ and the corresponding "required" value $(\cos\gamma)_r$. If $|(\cos\gamma)_r|\leqslant1$ and the solution $(\tilde{h}(\eta),\ \tilde{\theta}(\eta))$ satisfies the boundary conditions, we have solved the original problem. If the angle γ is not constrained, we may take $\cos\gamma\equiv(\cos\gamma)_r$. If $\cos\gamma$ is constrained to the two values ±1, $\tilde{h}(\eta),\tilde{\theta}(\eta)$ may be approximated in D by constructing a sliding program with the basis control functions $(\tilde{c}_{x0}(\eta),\ \tilde{c}_y(\eta),\ \cos\gamma_{1,\ 2})$, where $\cos\gamma_1=-1,\cos\gamma_2=+1$.

§ 5.2.   OPTIMAL DESCENT PROGRAM OF AN AIRCRAFT IN THE ATMOSPHERE BASED ON THE CONDITION OF MINIMUM ADDED INTEGRAL HEAT IN THE CRITICAL ZONE

5.2.1.   Statement of the problem

Consider the problem of aircraft descent in the atmosphere with minimum added heat in the critical zone.   The added heat is estimated by the functional

$$I = \int\limits_0^{t_1} K_q \varrho^{0.5} V^{3.15}\, dt,$$
(5.41)

where K_q is a constant.

The admissible motion programs satisfy the following conditions.

The equations of motion

$$\dot{h} = V \sin \theta;$$
(5.42)

$$\dot{V} = -X(h, V, c_y) - g \sin \theta;$$
(5.43)

$$\dot{\theta} = -\frac{1}{V}\left[Y_a + \left(\frac{V^2}{r_E + h} - g \right) \cos \theta \right],$$
(5.44)

where

$$Y_a = Y(h, V, c_y) \cos \gamma.$$

Boundary conditions:

A l t e r n a t i v e 1.   Only the initial and the final energy levels are given:

$$t = 0, \quad \eta(0) = \eta_{0f}$$
$$t = t_1, \quad \eta(t_1) = \eta_1$$
(5.45)

(transition from one energy level to another).

A l t e r n a t i v e 2.   The initial values of the phase coordinates and the final value of the energy level are given:

$$t = 0, \quad h(0) = h_{0f}; \quad V(0) = V_{0f}; \quad \theta(0) = \theta_{0f}; \quad \eta_1 = \eta_{1f}.$$
(5.46)

In both alternatives, t_1 is free.

The sought optimal program is expected to satisfy the following constraints:

1) $c_y \leqslant c_{y\,\lim};$
(5.47)

2) $h \geqslant h_{\lim};$
(5.48)

3) $[N(h, V, c_y)]^2 \equiv \frac{1}{g^2}[X^2(h, V, c_y) + Y^2(h, V, c_y)] \leqslant N_{\lim}^2;$
(5.49)

4) $q_t(h, V, c_y) \leqslant q_{t.\lim}.$
(5.50)

Inequality (5.47) is an engineering constraint on the lift coefficient; $c_{y\,\text{lim}}$ may be identified with $c_{y\,\text{nil}}$, say. In (5.48) h_{lim} is the geometrical altitude limit, i. e., the ground level. Relations (5.49), (5.50) impose constraints on the total overload N and the heat flux q_t at some critical point on the aircraft surface. The limit values of these parameters are fixed from structural strength considerations and from safety requirements for the crew and the instrumentation. It is assumed that for any fixed h and V, $q_t(h, V, c_y)$ is a symmetrical and monotonic (on both sides of the origin) function of c_y.

From considerations of problem solution, constrainsts (5.49), (5.50) are better written in the form (5.5). Note that both N^2 and q_t for any fixed h, V have a minimum for $c_y = 0$.

Condition (5.49) is then equivalent to the following two inequalities:

$$(N^2)_{\alpha=0} \leqslant N^2_{\text{lim}}; \tag{5.51}$$

$$|c_y| \leqslant c_y^*, \tag{5.52}$$

where $c_y^*(h, V)$ is the positive solution of the equation

$$N^2(h, V, c_y) = N^2_{\text{lim}} \tag{5.53}$$

for any fixed h and V which satisfy (5.51). Clearly, for other h, V, the solution c_y^* does not exist.

Condition (5.50) is similarly equivalent to the following inequalities:

$$(q_t)_{c_y=0} \leqslant q_{t.\text{lim}}; \tag{5.54}$$

$$c_y \leqslant c_y^{**}(h, V), \tag{5.55}$$

where $c_y^{**}(h, V)$ is the positive solution of the equation

$$q_t(h, V, c_y) = q_{t.\text{lim}}. \tag{5.56}$$

The functions $(N^2(h, V))_{c_y=0}$ and $(q_t(h, V))_{c_y=0}$ decrease monotonically with increasing altitude and increase monotonically with increasing velocity. Therefore (5.51) may be written in the form

$$h \geqslant h^*(V), \tag{5.57}$$

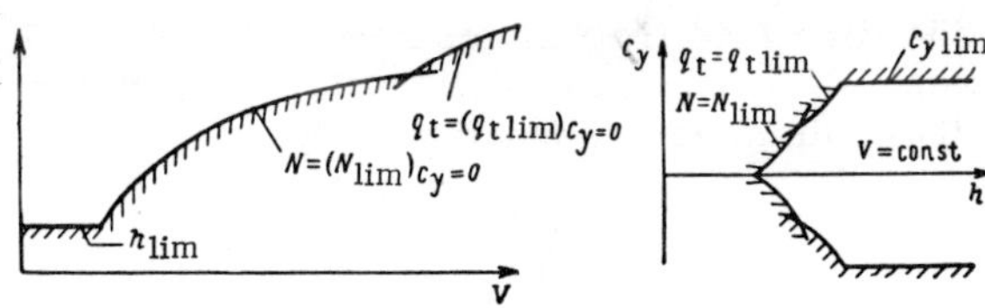

FIGURE 5.1

where $h^*(V)$ is the solution of the equation

$$(N^2(h, V))_{c_y=0} = N^2_{\text{lim}} \tag{5.58}$$

for any fixed V, and (5.54) may be written in the form

$$h \geqslant h^{**}(V), \tag{5.59}$$

where $h^{**}(V)$ is the solution of the equation

$$(q_t(h, V))_{c_y=0} = q_{t.\text{lim}} \tag{5.60}$$

for any fixed V.

Summing up, we can write the various constraints in the following form (Figure 5.1):

$$h \geqslant h_b(V) = \begin{cases} h_{\overline{\text{lim}}} & \text{for} \quad h_{\overline{\text{lim}}} > h^*, \ h^{**} \\ h^*(V) & \text{for} \quad h^* > h_{\overline{\text{lim}}}, \ h^{**} \\ h^{**}(V) & \text{for} \quad h^{**} > h_{\overline{\text{lim}}}, \ h^* \end{cases} \tag{5.61}$$

$$|c_y| \leqslant c_{yb}(h, V) = \begin{cases} c_{y\overline{\text{lim}}} & \text{for} \quad c_{y\overline{\text{lim}}} < c_y^*, \ c_y^{**} \\ c_y^*(h, V) & \text{for} \quad c_y^* < c_{y\overline{\text{lim}}}, \ c_y^{**} \\ c_y^{**}(h, V) & \text{for} \quad c_y^{**} < c_{y\text{lim}}, \ c_y^* \end{cases} \tag{5.62}$$

$$-Y(h, V, c_{yb}) \leqslant Y_a \leqslant Y(h, V, c_{yb}), \tag{5.63}$$

and in this form they will be used in what follows.

Note that the physical meaning of the constraints and the corresponding classification are of no consequence from the point of view of the method of problem solution. We are merely concerned with the formal classification of constraints, which distinguishes between constraints on the phase coordinates (group (5.61)) and constraints on the control elements depending on the phase coordinates (group (5.62)).

We can now present a rigorous formulation of the problem, denoting by D the set of admissible motion programs satisfying the above conditions:

Find a sequence of programs $\{(\bar{h}_s(t), \bar{V}_s(t), \bar{\theta}_s(t), \bar{c}_{ys}(t), \bar{\gamma}_s(t))\} \subset D$ on which the functional (5.41) goes to its least value over the set D.

5.2.2. Solution of the problem

Seeing that the time t_1 is free and constraints (5.61)–(5.63) do not depend explicitly on time, we will solve Problem 3 corresponding to the original problem. We thus minimize the functional

$$I = -K_q \int_{\eta_0}^{\eta_1} \frac{\varrho^{0.5} V^{3.15}}{\dfrac{V}{g} X(h, V, c_y)} \, d\eta, \tag{5.64}$$

where

$$V = \sqrt{2g(\eta - h)}$$

under constraints (5.61), (5.62). The minimum of the functional (5.64) yields a certain solution $\tilde{h}(\eta)$, $\tilde{c}_y(\eta)$ on which, for every fixed $\eta \in (\eta_1, \eta_0)$, the integrand in (5.64) attains its minimum under the same constraints.

Since in our case $f^0(h, V) > 0$, the minimum of the integrand for any fixed η and h is attained for $|c_y| = c_{yb}(h, \eta)$. We now insert $|c_y| = c_{yb}$ in (5.64) and find the minimum of the function

$$\varkappa(\eta, h) = \frac{\varrho^{0.5}V^{3.15}}{\dfrac{V}{g}X(h, V, c_{yb}(h, \eta))} \tag{5.65}$$

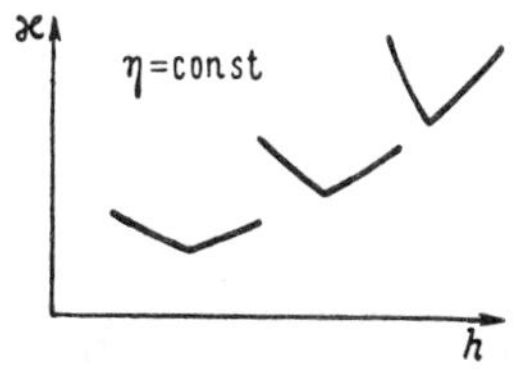

FIGURE 5.2

on the set $h \geqslant h_b(\eta)$. Here $h_b(\eta)$ is obtained by transforming $h_b(V)$ to the new coordinates (see Figure 5.1).

Figure 5.2 shows some typical curves of the function $\varkappa(h)$ for several fixed values of η. The minimum of $\varkappa(h)$ is attained at the kink point $\tilde{h}_{opt}$, which is precisely the point of transition from one c_y, c_{ylim}, to another c_y^* or c_y^{**}. In other words, this is the point where

$$N(h, \eta, c_{y\,lim}) = N_{lim}$$

or

$$q_t(h, \eta, c_{y\,\overline{lim}}) = \overline{q}_{t.lim}.$$

The solution of Problem 3 for the two alternative sets of boundary conditions thus has the following form (Figure 5.3).

Alternative 1.

$$\tilde{h}(\eta) = \begin{cases} \tilde{h}_{opt}(\eta) & \text{for} \quad h_{opt}(\eta) \geqslant \overline{h}_{lim} \\ \overline{h}_{lim} & \text{for} \quad h_{opt} < \overline{h}_{lim}. \end{cases} \tag{5.66}$$

$$\eta \in [\eta_1, \eta_0]$$

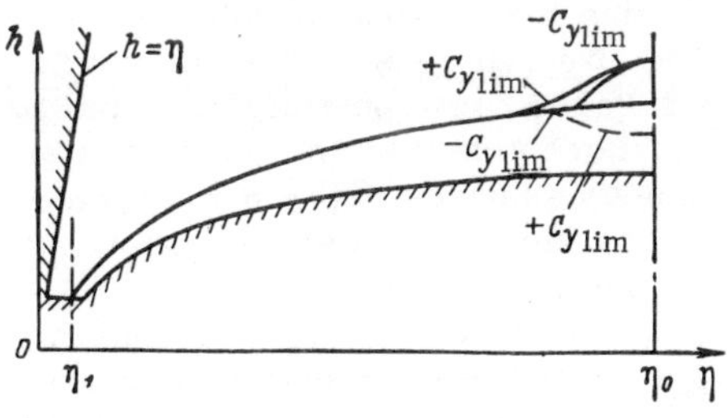

FIGURE 5.3

135

Alternative 2.

$$h=\begin{cases} h=h_{0f} & \text{for} \quad \eta=\eta_0 \\ \tilde{h}_{\text{opt}}(h) & \text{for} \quad \tilde{h}_{\text{opt}}(\eta) \geqslant h_{\text{lim}}. \\ h_{\text{lim}} & \text{for} \quad h_{\text{opt}} < h_{\text{lim}}. \end{cases} \quad \eta \in (\eta_1, \eta_0) \qquad (5.67)$$

For both alternatives,

$$|c_y| = c_{y\,\text{lim}}.$$

We will now consider the existence of the solution of Problem 3 in D, i. e., the actual construction of the sought minimizing solution.

In equations $(5.42)-(5.44)$ we change over to a new argument η:

$$h' = -\frac{g \sin \theta}{X(h, V, c_y)} = f^1(\eta, h, \theta, c_y); \qquad (5.68)$$

$$\theta' = \frac{g\left[Y_a + \left(\dfrac{V^2}{r_E} - g\right) \cos \theta \right]}{V^2 X(h, V, c_y)} = f^2(\eta, h, \theta, c_y). \qquad (5.69)$$

The solution of Problem 3 is feasible in D only if the direction $(1, \tilde{\theta}'(\eta), \tilde{h}(\eta))$ in the (h, θ, η) space, for every η, lies within the angle ω spanned by the vectors

$$a_1 = [1, f(\eta, \tilde{h}, \tilde{\theta}, c_{y\,\text{lim}})]$$

and

$$a_2 = [1, f(\eta, \tilde{h}, \tilde{\theta}, -c_y],$$

where $f = (f^1, f^2)$.

In other words, the substitution of $\bar{\theta}'(\eta)$ in the equation

$$\theta'(\eta) = v f^2(\eta, \tilde{h}, \tilde{\theta}, -c_{y\,\text{lim}}) + (1-v) f^2(\eta, \tilde{h}, \tilde{\theta} + c_{y\,\text{lim}}) \qquad (5.70)$$

should give v which satisfies the constraints $0 \leqslant v \leqslant 1$.

At the same time, the functions $\tilde{h}(\eta)$, $\tilde{\theta}(\eta)$ may be discontinuous. The values of $\tilde{h}'$, $\tilde{\theta}'$ at the discontinuity points are not bounded, and the direction $[1, \tilde{h}', \tilde{\theta}']$ therefore a priori does not lie within the angle ω since the functions $f^2(h, \theta, \eta, \pm c_{y\,\text{lim}})$ are bounded. In this sense, the solution of Problem 3 does not solve at the same time the original problem. However, using this solution as a guide line, we can construct some program $\bar{h}(\eta)$, $\bar{\theta}(\eta)$, $\bar{c}_y(\eta)$ from D which can be regarded as a potential optimal program.

Note that under the above constraints, the condition $0 \leqslant v \leqslant 1$ is generally satisfied on the continuous sections $(\tilde{h}(\eta), \tilde{\theta}(\eta))$ for a wide range of η, with the exception of very small η, when motion along $\tilde{h}_{\text{opt}}(\eta)$ requires smaller values of Y_a than the $Y_{a_{\text{lim}}}$ available.

A potential optimal program constructed in this way is shown in general outline in Figure 5.3. It comprises flight sections corresponding to the limiting values of Y_a and sliding control sections, with the control switching between the two limit values of Y_a along the $\hbar(\eta)$ line. In particular cases, some of the sections shown in Figure 5.3 may be missing.

5.2.3. Estimates of the solution

Since the solution $\bar{h}(\eta)$, $\bar{\theta}(\eta)$, $\bar{c}_y(\eta)$ constructed above is only a potential optimal solution, it requires further verification. One of the possible techniques is to verify the sufficient conditions of a local minimum /1/. Another approach is to improve the solution by successive approximations (e. g., by approximate optimal synthesis), using the available solution as the first approximation (a reference solution).

We propose to check the potential optimal solution by comparison with another solution $(h(\eta), \theta(\eta), c_y(\eta))_c$ which does not necessarily belong to D but on which the functional is a priori known to attain a value not exceeding its exact lower bound on D.

Let

$$(h(\eta), \theta(\eta), c_y(\eta)) = z.$$

Then

$$I(z_c) \leqslant \inf_D I. \tag{5.71}$$

By (5.71) we clearly have

$$|\Delta I| = |I(\bar{z}) - \inf_D I| \leqslant |I(\bar{z}) - I(z_c)| = \Delta I_c.$$

Thus ΔI_c provides an estimate of the solution, and the accuracy of this estimate clearly depends on the particular function z_c chosen.

We can choose z_c, say, as the solution of Problem 3. For boundary conditions in Alternative 1, this estimate is expected to be fairly accurate. For boundary conditions in Alternative 2, on the other hand, this estimate may prove to be too crude, since in this case the potential minimizing solution may markedly differ from z_c over some long sections (see Figure 5.3).

A more exact estimate can be obtained, but this requires some additional construction.

Using initial conditions (5.46), let us consider the solution of the system of six equations (with the argument $Q = \int_0^t \varrho^{0.5} V^{3.15} dt$)

$$\left.\begin{array}{l}
h'_L = \min\limits_{c_y, \eta, \theta} \left(\dfrac{V \sin \theta}{\varrho^{0.5} V^{3.15}} \right); \\[2em]
\eta'_L = \min\limits_{c_y, h, \theta} \left(-\dfrac{V}{g} \ \dfrac{X(V, h, c_y)}{\varrho^{0.5} V^{3.15}} \right); \\[2em]
\theta'_L = \min\limits_{Y_a, c_y, \eta, h} \left(\dfrac{1}{V} \ \dfrac{Y_c + \left(\dfrac{V^2}{r_E} - g \right) \cos \theta}{\varrho^{0.5} V^{3.15}} \right); \\[3em]
h'_U = \max\limits_{c_y, \eta, \theta} \left(\dfrac{V \sin \theta}{\varrho^{0.5} V^{3.15}} \right) \\[2em]
\cdots \cdots \cdots \cdots \cdots \cdots \cdots \cdots
\end{array}\right\} \tag{5.72}$$

The minima and the maxima of the corresponding functions are located for every fixed $\bar{Q}$ in the region

$$
\left.
\begin{array}{c}
-c_{y\lim}(h,\,\eta)\leqslant c_y\leqslant c_{y\lim}(h,\,\eta);\\[4pt]
-Y(h,\,\eta,\,c_{y_{\lim}})\leqslant Y_a\leqslant Y(h,\,\eta,\,c_{y\lim});\\[4pt]
h_{\mathrm{L}}(Q)\leqslant h\leqslant h_{\mathrm{U}}(Q);\\[4pt]
\eta_{\mathrm{L}}(Q)\leqslant\eta\leqslant\eta_{\mathrm{U}}(Q);\\[4pt]
\theta_{\mathrm{L}}(Q)\leqslant\theta\leqslant\theta_{\mathrm{U}}(Q).
\end{array}
\right\}
\qquad (5.73)
$$

In virtue of the right-hand sides of equations (5.72) (see Appendix), $\eta_{\mathrm{L}}(\bar{Q})$ is some lower limit value of η (which lies below the exact lower limit of η for the original system) for every fixed Q, and its inverse function is some lower limit of Q for fixed η (Figure 5.4). We continue the solution of this system to its intersection in the plane $(h_{\mathrm{L}},\,\eta_{\mathrm{L}})$ with the line $\hbar(\eta),\eta$. The quantities relating to the intersection point are denoted with an asterisk $*$. A similar construction can be carried out at the left end point of the interval $(\eta_1,\,\eta_0)$ until $h_{\mathrm{L}}(\eta_{\mathrm{L}})$ intersects $\hbar(\eta)$ at the point η^{**}.

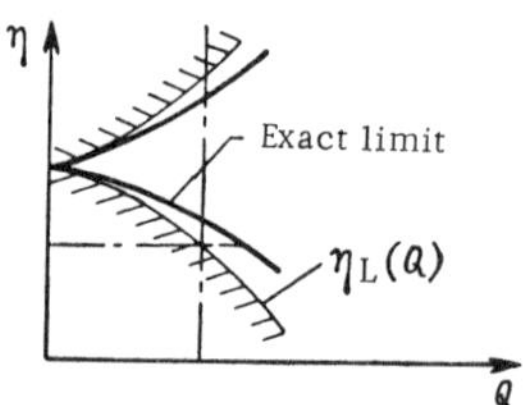

FIGURE 5.4

I_{c} is taken in the form

$$
I_{\mathrm{c}}=K_q\left(\Delta Q^{*}+\Delta Q^{**}+\int_{\eta^{**}}^{\eta^{*}}\frac{\varrho^{0.5}\widetilde{V}^{3.15}}{\dfrac{V}{g}X(\widetilde{h},\,\widetilde{V},\,c_{y\lim})}\,d\eta\right).
\qquad (5.74)
$$

The estimate is obtained as follows:

$$
\Delta I_{\mathrm{c}}=\left|I_{\mathrm{c}}-I(\bar{z})\right|
\qquad (5.75)
$$

or

$$
\bar{\Delta}=\frac{\Delta I_{\mathrm{c}}}{I(\bar{z})}.
\qquad (5.75a)
$$

5.2.4. Numerical example

As an example, we computed the optimal descent of a hypothetical aircraft with the following specifications:

$$\text{weight } G = 5000 \text{ km, wing load } \frac{G}{S} = 153 \text{ kg}/\text{m}^2.$$

The polar curve of the aircraft is

$$\begin{aligned}
c_x &= 0.1 + 1.83 \, |\sin^3 \alpha|; \\
c_y &= \pm 1.83 \sin^3 \alpha \cos \alpha.
\end{aligned} \qquad (5.76)$$

(Expressions (5.76) are borrowed from /6/.)

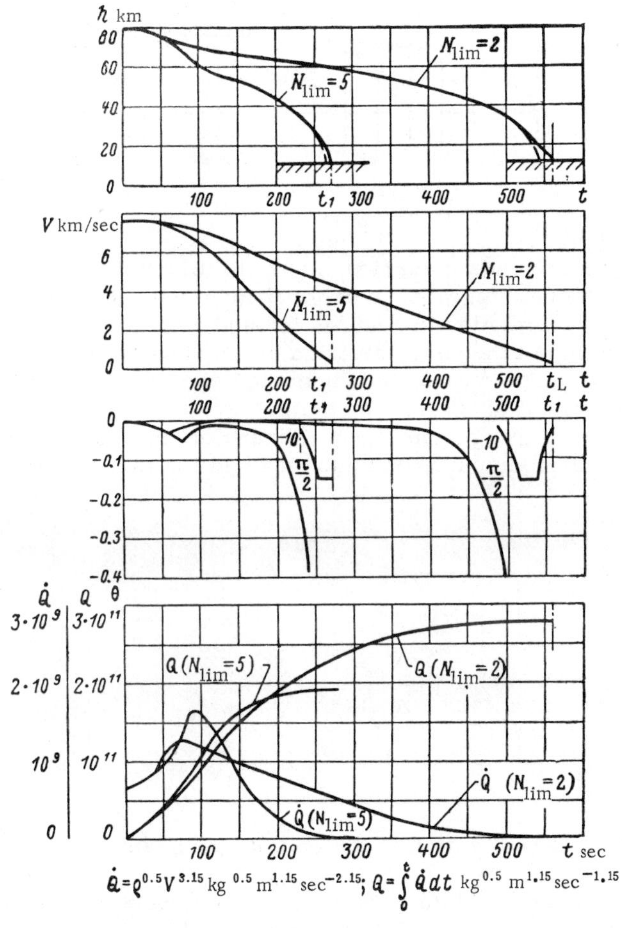

FIGURE 5.5

The following boundary conditions and constraints were used:

$$\eta_0 = 2.93 \cdot 10^6 \, \text{m}; \quad h_0 = 80 \cdot 10^3 \, \text{m}; \quad \theta_0 = 0;$$
$$\eta_1 = 15 \cdot 10^3 \, \text{m};$$
$$N_{\text{lim}} = 5; \ 2 \qquad c_{y\,\text{lim}} = 0.645. \tag{5.77}$$

No constraints were imposed on the heat flux. The density of the atmosphere was taken in accordance with the SA-64 standard atmosphere model.

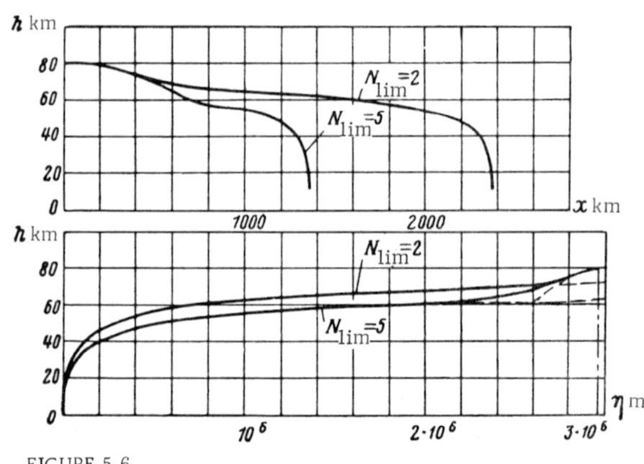

FIGURE 5.6

The computations were carried out on the BESM-2 computer. Equations $(5.42) - (5.44)$ were integrated in original form with the argument t.

The results are plotted in Figures 5.5 and 5.6.

Figure 5.5 shows the time variation of the phase coordinates h, V, θ and the variables $\dot{Q}$, Q for $N_{\text{lim}} = 5, 2$. Figure 5.6 shows the auxiliary characteristics of the optimal program: the trajectory in the coordinates h, x, where x is the horizontal range, and h, η. Figure 5.6 also gives the function $h_L(\eta_L)$. The following estimates are obtained for this program.

Alternative 1.

$$\overline{\Delta} \approx 0.$$

Alternative 2.

$$N_{\text{lim}} = 5; \quad \overline{\Delta} = 4.1\%;$$
$$N_{\text{lim}} = 2; \quad \overline{\Delta} = 0.6\%.$$

5.2.5. Practical realization of the optimal program

In the particular problem considered in this section, the limits for the variation of the angle of attack are assumed to be equal in absolute value.

The sought solution in this case can be realized in more than one way. Let us consider in some detail two realizations of the optimal program.

According to one technique, the angle of attack is controlled by altering the angle of pitch while maintaining a constant angle of bank $\gamma=0$ or $\gamma=\pi$. The sections with $Y_a=-Y_{\lim}$ or $Y_a=Y_{\lim}$ may be attained by taking $c_y=-c_{y\lim}$ or $c_y=c_{y\lim}$, respectively. Sections with $-Y_{\lim}<Y_a<Y_{\lim}$ in this case are sliding control sections and are attained by switching between the two limit values of the angle of attack.

According to another technique, the banking of the aircraft is controlled so that $c_y=c_{y\lim}$ or $c_y=-c_{y\lim}$. A number of alternative versions of this technique may be considered, corresponding to different programs of al altering the angle of bank. All these programs are characterized by the same "mean" projection of the lift on the vertical plane, which is equal to Y_a, whereas the "mean" projection of the lift on the horizontal plane is variable. One of the possible programs may maintain the instantaneous angle of bank in accordance with the relation

$$\cos\gamma=\frac{Y_a}{Y_{\lim}}, \tag{5.78}$$

another program may rotate the aircraft with a variable angular banking velocity, a third program may induce oscillatory movement of the aircraft in a certain range of banking angles, etc. This freedom of banking control creates wide possibilities for lateral maneuvers.

With the first control technique, the sliding control section is realized by switching over between the two limit angles of attack in flight. An elementary switching cycle is shown in Figure 5.7. To assess the departure of the real values of the functional from its lower bound, let us consider the expression of the functional for $c_y=c_{yb}(h,\eta)$ The integrand $\varkappa(h,\eta)$ for every fixed η depends only on the altitude h. The deviations of the functional from its minimum are determined by the altitude deviations

$$\Delta h=h-\tilde{h}(\eta)$$

for all

$$\eta\in(\eta_1,\eta_0).$$

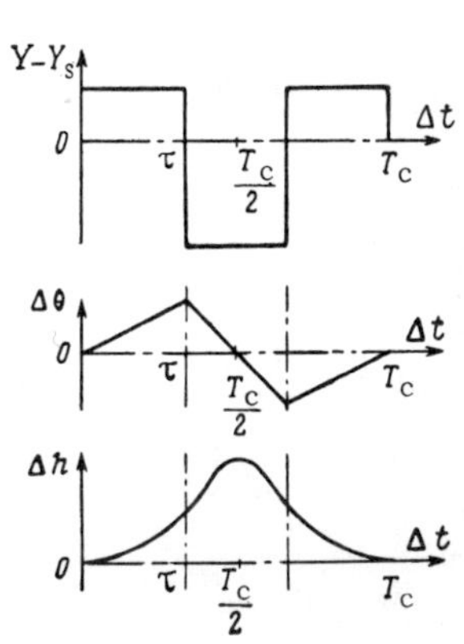

FIGURE 5.7

Let us estimate the magnitude of these deviations for different number of cycles. To this end, we introduce the following assumptions. In the range of altitudes corresponding to the permissible deviations Δh, the velocity V and the density ϱ are taken equal to the optimal values $\overline{V}(\eta)$, $\tilde{\varrho}(\eta)$; $\cos\theta\approx1$; $\sin\theta\approx\theta$. Then in the relevant range of altitudes, for fixed η,

$$X=\overline{X}=\text{const};$$
$$Y=\overline{Y}=\text{const}.$$

We further assume an exponential density distribution in the atmosphere,

$$\varrho = \varrho_0 e^{-\beta(h-h_0)}, \quad \beta = 1.31 \cdot 10^{-4} \, \mathrm{m}^{-1}.$$

Under these assumptions, the equations of the relative deviations $\Delta h = h - \tilde{h}$, $\Delta\theta = \theta - \tilde{\theta}$ take the form

$$\left.\begin{array}{c} (\Delta\dot{h}) = \tilde{V}\Delta\theta; \\[2mm] (\Delta\dot{\theta}) = \dfrac{1}{\tilde{V}}(\pm Y - Y_s), \end{array}\right\} \tag{5.79}$$

where

$$Y_s = v Y(\tilde{h}, \tilde{V}, \tilde{c}_{y\lim}) + (1-v) Y(\tilde{h}, \tilde{V} - \tilde{c}_{y\lim}).$$

The change in velocity during a single cycle is ignored.

The integrand $\varkappa(\eta, h)$ in (5.65) may be replaced in the relevant neighborhood of $\tilde{h}$ by the expression

$$\varkappa(\eta, h) = \tilde{\varkappa} + \tilde{\varkappa}_h \Delta h, \quad \Delta h \geqslant 0, \tag{5.80}$$

where, as is readily seen,

$$\left.\begin{array}{c} \tilde{\varkappa}_h = \tilde{\varkappa}\xi(\tilde{V}^2); \\[2mm] \xi(\tilde{V}^2) = a\beta - \dfrac{b}{\tilde{V}^2}; \; a = 0.5; b = 0.007. \end{array}\right\} \tag{5.81}$$

Hence we easily find the relative deviation of the functional from its minimum over a cycle

$$\frac{\delta I}{\Delta I} = \xi(\tilde{V}^2)\frac{2}{T_c} \int\limits_0^{T_c/2} \Delta h(t)\, dt. \tag{5.82}$$

Let us evaluate the integral in (5.82) using equations (5.79). Integration of (5.79) over Δt (see Figure 5.7) gives

$$\Delta h = (Y - Y_s)\frac{(\Delta t)^2}{2}, \quad 0 \leqslant \Delta t \leqslant \tau,$$

$$\Delta h = (Y - Y_s)\frac{\tau^2}{2} + (Y - Y_s)\tau(\Delta t - \tau) - (Y - Y_s)\frac{(\Delta t - \tau)^2}{2},$$

$$\tau < \Delta t < \frac{T_c}{2},$$

where

$$\tau = \frac{1}{4}\frac{Y + Y_s}{Y} T_c.$$

Inserting the expressions for Δh in (5.82) and integrating, we obtain

$$\frac{\delta I}{\Delta I} = \frac{1}{4}\xi(\tilde{V}^2)\left(\frac{T_c}{2}\right)^2 \left\{ \frac{1}{3}(Y - Y_s)\left(\frac{Y + Y_s}{Y}\right)^3 + \right.$$

$$+ (Y - Y_s)\left(\frac{Y - Y_s}{Y}\right)\left(\frac{Y + Y_s}{Y}\right)^2 - \frac{1}{3}(Y + Y_s)\left(\frac{Y - Y_s}{Y}\right)^3 +$$

$$+(Y-Y_s)\left(\frac{Y-Y_s}{Y}\right)^2\left(\frac{Y+Y_s}{Y}\right)\bigg\}.$$

T_c is related to the number of switchings by the equality

$$T_c=\frac{t_2-t_1}{2N}; \qquad N=\frac{t_2-t_1}{2T_c}.$$

The total deviation δI is thus expressed by the relation

$$\frac{\delta I}{I}=\frac{\sum\limits_{i=1}^{N}\left(\frac{\delta I}{\Delta I}\right)_i(\Delta I)_i}{I}=\sum\limits_{1}^{N}\left(\frac{\delta I}{\Delta I}\right)_i\frac{(\Delta I)_i}{I}. \tag{5.83}$$

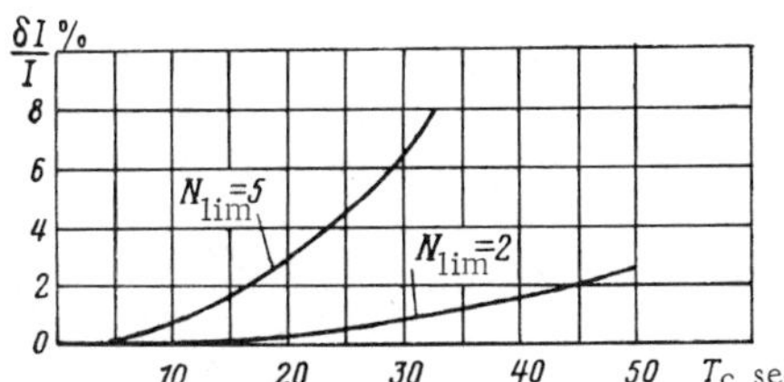

FIGURE 5.8

The dependence of $\delta I/I$ on T_c is shown in Figure 5.8. It is readily seen that a program with $T_c\approx 30$ sec is already sufficient to provide an adequate approximation to the optimum. The transition time between the two limit values of c_y apparently can be ignored in comparison with T_c.

The deviation of the functional from the true minimum in the second technique depends on the particular banking control program. It seems that, irrespective of the particular program, banking control invariably gives a better realization of the optimum for the same number of cycles, since in this case $|c_y|=c_{y\lim}$ at any time. The program fixing the angle of bank in accordance with (5.78) attains the exact lower bound of the functional.

5.2.6. Remarks concerning a controlled drag coefficient

We will now see how the solution of the problem is affected if c_{x0} is controllable within certain limits

$$c_{x0\,min}\leqslant c_{x0}\leqslant c_{x0\,max},$$

which are either imposed by engineering constraints or are functions of h, V, and c_y, as it follows from the condition

$$N(h,\,V,\,c_{x0},\,c_y)\leqslant N_{\lim} \tag{5.84}$$

(the constraint on the heat flux is disregarded).

Condition (5.84) is better written as a system of two constraints

$$N(h, V, c_{x0}, 0) \leqslant N_{\lim};\qquad\qquad\qquad (5.85)$$

$$c_x(c_{x0}, c_y) \leqslant c_{x\lim}(h, V),\qquad\qquad\qquad (5.86)$$

where $c_{x\lim}(h, V)$ is the solution of the equation

$$N^2(h, V, c_x(c_{x0}, c_y)) = N^2_{\lim}.\qquad\qquad\qquad (5.87)$$

Let us investigate the integrand in (5.64) under these constraints, taking c_{x0} to be variable. Since $f^0 > 0$, the minimum of the integrand with respect to c_{x0}, c_y corresponds to the maximum with respect to c_{x0}, c_y of the function

$$X(h, V, c_x(c_{x0}, c_y)),$$

which, for any fixed h, η, is attained on the maximum value of $c_x(c_{x0}, c_y)$, i. e.,

$$c_x(c_{x0}, c_y) = c_{x\lim}(h, \eta).$$

Equation (5.87) may be written in the form

$$c_x^2(c_x, c_y) + c_y^2 = \frac{N^2_{\lim}}{\left[\dfrac{S}{2G}\,(\varrho V^2)\right]^2}.\qquad\qquad\qquad (5.87a)$$

The function $c_x(c_{x0}, c_y)$ attains its maximum, under constraint (5.87a), for $c_{x0} = c_{x0\,\max}(h, V)$, where $c_{x0\,\max}$ is either the solution of equation (5.84) for any fixed h, V (in this case $c_y = 0$) or the upper engineering limit $c_{x0\,\max f}$ (in this case $|c_y|$ is obtained from (5.87a)).

Inserting the results for c_{x0} and c_y in the integrand in (5.64), we find that the function $\varkappa(\eta, h)$ attains its minimum on $h = \hbar(\eta)$, which is determined from the condition

$$N^2(h, V, c_{x0\,\max f}, c_{y\lim}) = N^2_{\lim}.\qquad\qquad\qquad (5.88)$$

We thus see that the optimal programs with controlled c_{x0} are essentially the same as with a fixed c_{x0}. If the fixed c_{x0} coincides with $c_{x0\,\min}$ for the case of controlled c_{x0}, the altitude $\hbar(\eta)$ is evidently greater than with fixed c_{x0}.

Appendix

Estimating the limits of the admissible phase coordinates for systems of differential equations

Consider a system of differential equations

$$
\left.
\begin{aligned}
\dot{y}^1 &= f^1(t,\, y^1,\, y^2,\, \ldots,\, y^n,\, u); \\
\dot{y}^2 &= f^2(t,\, y^1,\, y^2,\, \ldots,\, y^n,\, u); \\
&\; \cdots \cdots \cdots \cdots \cdots \cdots \\
\dot{y}^n &= f^n(t,\, y^1,\, y^2,\, \ldots,\, y^n,\, u),
\end{aligned}
\right\}
\tag{5.89}
$$

with given initial conditions

$$
y^1(0) = y_0^1, \quad y^2(0) = y_0^2, \quad \ldots, \quad y^n(0) = y_0^n.
\tag{5.90}
$$

The functions $f^i\,(t,\, y^1,\, y^2,\, \ldots,\, y^n,\, u)$ are assumed continuous and differentiable in all their arguments, $u(t)$ is an r-dimensional vector function (control function) whose values lie in some region $Q(t,\, y^1,\, y^2,\, \ldots,\, y^n)$ in the r-dimensional space.

We use (5.89) to form a new system of $2n$ differential equations

$$
\left.
\begin{aligned}
\dot{y}_L^1 &= \inf_{y^1,y^2,\ldots,y^n,u} f^1(t,\, y^1,\, y^2,\, \ldots,\, y^n,\, u); \\
\dot{y}_U^1 &= \sup_{y^2,y^3 \ldots y^n,u} f^1(t,\, y^1,\, y^2,\, \ldots,\, y^n,\, u); \\
\dot{y}_L^2 &= \inf_{y^1,y^2,\ldots,y^n,u} f^2(t,\, y^1,\, y^2,\, \ldots,\, y^n,\, u); \\
\dot{y}_U^2 &= \sup_{y^1,y^2,\ldots,y^n u} f^2(t,\, y^1,\, y^2,\, \ldots,\, y^n,\, u); \\
&\; \cdots \cdots \cdots \cdots \cdots \cdots \cdots
\end{aligned}
\right\}
\tag{5.91}
$$

The maxima and the minima of the corresponding functions lie in the region

$$
\left.
\begin{aligned}
y_L^1(t) &\leqslant y^1 \leqslant y_U^1(t); \\
y_L^2(t) &\leqslant y^2 \leqslant y_U^2(t); \\
&\; \cdots \cdots \cdots \cdots \\
y_L^n(t) &\leqslant y^n \leqslant y_U^n(t).
\end{aligned}
\right\}
\tag{5.92}
$$

It can be shown that the functions $y_U^i\,(t)$, $y_L^i\,(t)$ corresponding to the solution of system (5.91) with initial conditions (5.90) constitute certain upper and lower limits (not necessarily exact) of the functions $y^i\,(t)$ corresponding to all the admissible solutions of system (5.89) with the same initial conditions.

We will first examine the question of limits for the case of one first-order differential equation. The following proposition holds true.

Lemma. Consider a first-order differential equation

$$\dot{y}=f(t,\,y,\,u),\tag{5.93}$$

where u is an r-dimensional vector function (a control function) taking its values in some region $Q(t,\,y)$ of the r-dimensional space. The function $f(t,\,y,\,u)$ is assumed to be continuous and differentiable in all its arguments. The initial condition $y(0)=y_0$ is given. The solution $y_L(t)$ of the equation

$$\dot{y}_L=\inf_{u\in Q(t,y)}f(t,\,y_L,\,u)\tag{5.94}$$

with this initial condition provides the exact lower limit for all the admissible values of y and every fixed $t\in[0,\,t_1]$ for the solutions of equation (5.93) with the same initial condition.

Indeed, the necessary conditions for a minimum of $y(t_1)=y_1$ in the Mayer problem for equation (5.93) are

$$R_y=0\to\psi=-\psi f_y,\quad \psi=\varphi_y\,(t,\,y_L);\tag{5.95}$$

$$R\,(t,\,y_L,\,\bar{u})=\sup_{u\in Q(t,y)}R\,(t,\,y_L,\,u);\tag{5.96}$$

$$\Phi_{y_1}=0\to\psi(t_1)=-1,\tag{5.97}$$

where

$$R\,(t,\,y,\,u)=\varphi_y f\,(t,\,y,\,u)+\varphi_t;\tag{5.98}$$

$$\Phi\,(t_1,\,y_1)=y_1+\varphi(t_1,\,y_1);\tag{5.99}$$

$\varphi(t,\,y)$ is an arbitrary continuous and differentiable function.

It follows from (5.95), (5.97) that $\psi(t)<0$ everywhere on $(0,\,t_1)$, being a solution of a homogeneous linear equation which is negative at least at one point. But then (5.96) is equivalent to (5.95), which together with the initial condition $y(0)=y_0$ defines a unique solution of the equation. Since (5.95) — (5.97) are necessary conditions, the unique solution is of necessity optimal. Thus $y_L(t_1)$ is the absolute minimum or the exact lower limit of $y(t_1)$ for equation (5.93) with the given initial condition. Since $t_1>0$ is arbitrary time, $y_L(t)$ is the exact lower limit of all the admissible values of y for any t.

It is similarly proved that the solution of the equation

$$\dot{y}_L=\sup_{u\in Q(t,y)}f\,(t,\,y,\,u)\tag{5.100}$$

with the same conditions is the exact upper bound of y for every fixed $t>0$.

Let us now return to systems (5.89), (5.91). If each equation from (5.89), say the equation for y^1, is considered independently of the other equations, it has the same form as equation (5.93), with $u(t)$ and the phase coordinates except y^1 treated as the control functions. The first and the second equation in (5.91) have the form of equations (5.94) and (5.100) in

146

relation to (5.93) and they therefore determine the upper and the lower limit of y^1 if the region of the values of y^i where the maximum and the minimum are sought coincides with the region of the admissible values of y^i or encloses it.

The initial region of the coordinates is given: in our case this is the point $y^i(0) = y_0^i$. At every succeeding time, equations (5.91) define the limits of the admissible values of y^i, so that (5.92) define the region of the admissible values of the variables at any time, etc.

Hence it follows that the solution of system (5.91) defines the limits of the components of the solution of system (5.89) at any time.

Bibliography for Chapter V

1. Krotov, V.F. and V.I. Gurman. Optimal'noe regulirovanie ugla ataki krylatogo apparata iz usloviya minimuma konechnoi skorosti (Optimum Angle-of-Attack Control of a Winged Aircraft under the Condition of Minimum Final Velocity). — In: "Issledovaniya po dinamike poleta," Mashinostroenie. 1969.
2. Rulev, V.A. O neobkhodimykh i dostatochnykh usloviyakh ekstremuma v variatsionnykh zadachakh dinamiki poleta letatel'nykh apparatov (Necessary and Sufficient Extremum Conditions in Variation Problems of Aircraft Flight Dynamics). — Izvestiya Vuzov, aviation engng, No.1. 1967.
3. Sivkov, G.V. Naivygodneishie manevry v vertikal'noi ploskosti apparata peremennoi massy (Optimal Vertical-Plane Maneuvers of a Varying Mass Vehicle). — Izvestiya Vuzov, aviation engng, No.4. 1959.
4. Tarasov, E.V. Optimal'nye rezhimy poleta letatel'nykh apparatov (Optimal Aircraft Flight Policies). — Oborongiz. 1963.
5. Breakwell, J.V. The Optimization of Trajectories. — J. Soc. Ind. Appl. Math., No.7. 1959.
6. Dynasoar. — Aero/Space Engng, 20, No.8. 1960

Chapter VI

NECESSARY AND SUFFICIENT OPTIMUM CONDITIONS FOR DISCRETE CONTROLLED SYSTEMS

§ 6. 1. STATEMENT OF THE PROBLEM

Consider two quite general sets Y and U with elements y and u, respectively, and a finite natural sequence $A = (0, 1, 2, \ldots, N)$. To every $i \in A$ corresponds a subset $V(i)$ of the direct product $Y \times U$.

We further introduce a set D of the pairs $y(i)$, $u(i)$ of functions of the integral argument i defined over A, such that for all $i \in A$

$$(y(i),\ u(i)) \in V(i); \quad i = 0, 1, \ldots, N; \tag{6.1}$$

$$y(i+1) = f[i,\ y(i),\ u(i)]; \quad i = 0, 1, \ldots, N-1, \tag{6.2}$$

where the function (operator) $f(i, y, u)$ is defined over the direct product $A \times Y \times U$, mapping it onto the set Y. It is assumed that D is nonempty. An element y is said to be a state of the system or its phase state; u is the control. The former differs from the latter in that it enters the constraint equations (6.2) with different values of i.

We define a functional on D,

$$I = \sum_{i=0}^{N} f^0[i,\ y(i),\ u(i)], \tag{6.3}$$

where $f^0(i, y, u)$ is a functional defined on $A \times Y \times U$. Let the functional I be bounded from below on D, i. e.,

$$\inf_{D} I = d > -\infty.$$

We seek a sequence $\{\bar{y}_S(i), \bar{u}_S(i)\} \subset D$ which minimizes the functional I over D, i. e., a sequence such that $I[\bar{y}_S(i), \bar{u}_S(i)] \to d$ for $S \to \infty$. In particular, if there exists an element $(\bar{y}(i), \bar{u}(i)) \in D$ satisfying the equality $I[\bar{y}(i), \bar{u}(i)] = d$ (this element is called the absolute minimizing solution), the problem reduces to the determination of this element.

We define an arbitrary functional $\varphi(i, y)$ on the direct product $A \times Y$ and use it in the following constructions:

$$R(i, y, u) = \varphi[i+1, f(i, y, u)] - \varphi(i, y) - f^0(i, y, u); \qquad (6.4)$$

$$\mu(i) = \sup_{(y,u) \in V(i)} R(i, y, u); \qquad (6.5)$$

$$\left.\begin{aligned}\Phi_0(y, u) &= -\varphi[1, f(0, y, u)] + f^0(0, y, u); \\ \Phi_1(y, u) &= \varphi(N, y) + f^0(N, y, u);\end{aligned}\right\} \qquad (6.6)$$

$$m_0 = \inf_{(y,u) \in V(0)} \Phi_0(y, u); \quad m_1 = \inf_{(y,u) \in V(N)} \Phi_1(y, u). \qquad (6.7)$$

We will prove the following theorem.

Theorem 6.1. (Optimum principle). Consider a sequence $\{\bar{y}_s(i), \bar{u}_s(i)\} \subset D$. For this sequence to minimize the functional I over D it is sufficient, and if for all $i \in A$ the functional $f^0(i, y, u)$ is bounded on $V(i)$ for every $i \in A$, it is also necessary that there exists a functional $\varphi(i, y)$ satisfying the following conditions:

1) a function $\mu(i)$ expressed by (6.5) is defined on $\{1, 2, \ldots, N-1\}$;
2) for all $i = 1, 2, \ldots, N-1,$

$$R[i, \bar{y}_s(i), \bar{u}_s(i)] \to \mu(i), \; S \to \infty; \qquad (6.8)$$

3)
$$\left.\begin{aligned}\Phi_0[\bar{y}_s(0), u_s(0)] &\to m_0, \\ \Phi_1[\bar{y}_s(N), \bar{u}_s(N)] &\to m_1, \; S \to \infty.\end{aligned}\right\} \qquad (6.9)$$

Remark. If the sequence introduced in the theorem has the form $\bar{y}_s(i) = \bar{y}(i), \bar{u}_s(i) = \bar{u}(i)$ for all S, the condition of convergence in (6.8), (6.9) is replaced by equality and the pair $\bar{y}(i), \bar{u}(i) \in D$ satisfying the conditions of the theorem is the absolute minimizing solution.

Sufficiency. We will use the lemma from § 1.2. The set M of this lemma is identified with D, and the set N is identified with a new set E which satisfies all the conditions of D except equalities (6.2). On this set we define a functional

$$L[y(i), u(i)] = \Phi_0(y(0), u(0)) + \Phi_1(y(N), u(N)) - \sum_{i=1}^{N-1} R[i, y(i), u(i)]. \qquad (6.10)$$

For $y(i), u(i) \in D$, $L = I$. This is quite obvious if we rewrite (6.10) in the form

$$L[y(i), u(i)] = I + \sum_{i=1}^{N} \varphi[i, y(i)] - \varphi[i, f(i-1), y(i-1), u(i-1)]. \qquad (6.11)$$

Suppose that there exists a functional $\varphi(i, y)$ such that the conditions of the theorem are satisfied on some sequence $\{y_s(i), u_s(i)\} \subset D$. This sequence then minimizes the functional L on E and, in virtue of the lemma, it also minimizes the functional I on D. The second part of the theorem —
— necessity — will be proved in § 6.5.

The present theorem generalizes to the discrete case the sufficient optimum conditions formulated for continuous processes. If the conditions of § 1.1 − § 1.2 are imposed on the sets Y and U and on the functional φ and constraint (6.2) is represented as a difference scheme

$$y(i+1) - y(i) = \Delta \widetilde{f}[i,\ y(i),\ u(i)],$$

where Δ is a positive number, we readily see that the functions R and Φ of the theorem coincide with the corresponding functions of § 1.2, apart from a factor Δ and terms of higher order in Δ. It is significant that the transition to an integral argument made it possible to reduce to a minimum the various mathematical concepts and constraints required for formulating the result. The problem is now stated in terms of general sets, operators defined on these sets, and functionals.

The optimum principle formulated in this section can be applied to reduce the problem of minimization of the functional I on D to the problem of maximizing the functional $R(i,\ y,\ u)$ on $V(i)$ for every $i = 1,\ 2,\ \ldots,\ N-1$ and the functionals $\Phi_0(y,\ u)$ and $\Phi_1(y,\ u)$. The necessary relation between these subproblems is established by an appropriate choice of the functional $\varphi(i,\ y)$.

The conditions of the theorem leave us considerable freedom in choosing the functional $\varphi(i,\ y)$. By imposing additional constraints on φ, we can thus develop various methods of solution within the framework of the proposed formalism, including discrete analogs of the continuous methods considered in previous chapters. We will now describe some of these analogs.

§ 6.3. BELLMAN'S METHOD

Suppose that for all i the set $V(i)$ coincides with the section $V^*(i)$ of the class D for a given i. Let $V_y(i)$ be the projection of V_i on Y, i. e., the set of elements $y \in Y$ each of which can be paired with at least one element u such that $(y,\ u) \in V(i)$; $V_u(i,\ y)$ is the section of $V(i)$ for a given $y \in V_y(i)$. On $V_y(i)$ we construct the functionals

$$P(i,\ y) = \sup_{u \in V_u(i,y)} R(i,\ y,\ u), \quad i = 1,\ 2, \ldots,\ N-1;$$

$$F_0(y) = \inf_{u \in V_u(N,y)} \Phi_0(y,\ u); \tag{6.12}$$

$$F_1(y) = \inf_{v \in V_u(N,y)} \Phi_1(y,\ u), \quad i = 0,\ 1, \ldots,\ N$$

and choose $\varphi(i,\ y)$ so that

1) the functional $P(i,\ y)$ exists and is independent of y,

$$P(i,\ y) = c(i),\ i = 1,\ 2,\ \ldots,\ N-1, \tag{6.13}$$

where $c(i)$ is an arbitrary function;

2) the functional $F_1(y)$ exists and is independent of y,

$$F_1(y) = c_1 \qquad (6.14)$$

where c_1 is an arbitrary number.

In practice, the selection of this φ amounts to solving a Cauchy problem for the functional equation (6.13) with the initial condition (6.14) in the direction from N to 0. Indeed, from (6.14) we find

$$\varphi(N, y) = -\inf_{u \in V_u(N, y)} f^0(N, y, u) + c_1,$$

and then from (6.13) for $i = N - 1$

$$\varphi(N - 1, y) = \sup_{u \in V_u(N-1, y)} \{\varphi[N, f(N-1, y, u)] - f^0(N-1, y, u)\}, \qquad (6.15)$$

etc. If the functional $f^0(i, y, u)$ is bounded on $V(i)$ for every $i \in A$, it is readily seen that the solution $\varphi(i, y)$ of this problem always exists.

Let $\{\bar{u}_s(i, y)\}$ be the sequence of the elements $u \in V_u(i, y)$ on which

$$R[i, y, \bar{u}_s(y, i)] \to P(i, y) \text{ for } i = 1, 2, \ldots, N-1 \qquad (6.16)$$

and

$$\left. \begin{array}{l} \Phi_0[y, \bar{u}_s(0, y)] \to F_0(y); \\ \Phi_1[y, \bar{u}_s(N, y)] \to F_1(y), \end{array} \right\} \qquad (6.17)$$
$$\scriptstyle s \to \infty$$

and let $\{\bar{y}_s(i)\}$ be the sequence of the solutions of the system

$$\left. \begin{array}{l} y(i+1) = f[i, y(i), \bar{u}_s(i, y(i))], \\ i = 0, 1, \ldots, N-1; \\ F_0[y_s(0)] \to m_0. \end{array} \right\} \qquad (6.18)$$

The sequence $\{\bar{y}_s(i), \bar{u}_s(i) = \bar{u}_s[i, y_s(i)]\}$ belongs to D and satisfies the sufficient conditions of the theorem, i. e., it is a minimizing sequence. This method of selecting $\varphi(i, y)$ thus leads to a complete solution of the problem.

If we take $c(i) \equiv 0$, and interpret $\varphi(i, y)$ as minus the "gain function", setting

$$\bar{y}(0) = y_0; \quad \bar{y}(N) = y_N,$$

the functional equation (6.13) coincides with the equation of Bellman's optimum principle /1/.

§ 6. 4. THE LAGRANGE—PONTRYAGIN METHOD

Let Y and U be finite-dimensional Euclidean spaces with the elements $y = (y^1, \ldots, y^n)$ and $u = (u^1, \ldots, u^r)$, respectively; $V(i) = V_y(i) \times V_u(i)$; the sets

$V_y(0)$ and $V_y(N)$ are the fixed points $y_0 \in Y$ and $y_N \in Y$, and $V_y(i)$ with $i=1, 2, \ldots, N-1$ coincide with Y; $V_u(i)$, $i=0, 1, \ldots, N$, coincide with U; the vector functions $f(i, y, u)$ and the functions $f^0(i, y, u)$ are continuous and differentiable on $V(i)$, $i=0, 1, \ldots, N$.

In this method, the function $\varphi(i, y)$ is sought simultaneously with the extremal $(\bar{y}(i), \bar{u}(i)) \in D$. Assuming $\varphi(i, y)$ to be continuous and differentiable with respect to y for every i and introducing the vector function $\psi(i)$, defined as the gradient of φ at the points of the extremal,

$$\psi(i) = \partial \varphi(i, \bar{y}) / \partial y |_{y = \bar{y}(i)}, \tag{6.19}$$

we write the necessary conditions for a maximum of R:

$$\left. \frac{\partial R}{\partial y} \right|_{y=\bar{y}(i), u=\bar{u}(i)} \equiv -\psi(i) + \frac{\partial}{\partial y} H[i, \psi(i+1), \bar{y}(i), \bar{u}(i)] = 0; \tag{6.20}$$

$$\left. \frac{\partial}{\partial u} R(i, y, u) \right|_{y=\bar{y}(i), u=\bar{u}(i)} \equiv \frac{\partial}{\partial u} H[i, \psi(i+1), \bar{y}(i), \bar{u}(i)] = 0, \tag{6.21}$$

where

$$H[i, \psi, y, u] = \psi f(i, y, u) - f^0(i, y, u). \tag{6.22}$$

These equations are a discrete analog of the Euler–Lagrange equations of variational calculus in Pontryagin's form. Together with the boundary condition

$$\left. \begin{aligned} &\bar{y}(0) = y_0; \quad \bar{y}(N) = y_N; \\ &\frac{\partial}{\partial u} \Phi_0[y_0, \bar{u}(0)] \equiv -\psi(1) \frac{\partial}{\partial u} f(0, y_0, \bar{u}(0)) + \\ &\qquad + \frac{\partial}{\partial u} f^0[0, y_0, \bar{u}(0)] = 0; \\ &\frac{\partial}{\partial u} \Phi_1[y_N, \bar{u}(N)] \equiv \frac{\partial}{\partial u} f^0[N, y_N, \bar{u}(N)] = 0, \end{aligned} \right\} \tag{6.23}$$

they define the extremal $(\bar{y}(i), \bar{u}(i)) \in D$ and $\psi(i)$. To bring the solution to completion, i.e., to prove that the pair $\bar{y}(i)$, $\bar{u}(i)$ is indeed the sought absolute minimizing solution, we must show that there exists a function $\varphi(i, y)$ satisfying (6.8), (6.9), and (6.19).

R e m a r k. Equations (6.21) are the necessary conditions for the maximum of $R(i, y, u)$ and also for the minimum of I on $D/3/$. Further note that in the discrete case, as distinct from the continuous case, the maximum of the Hamiltonian H with respect to u (Pontryagin's maximum principle) does not provide a necessary condition for the maximum of $R(i, y, u)$, nor is it a necessary condition of optimum $/3/$.

§ 6.5. PROOF OF NECESSITY OF THE CONDITIONS OF THE OPTIMUM PRINCIPLE

We will now prove the second part of Theorem 6.1, the n e c e s s i t y. It follows from the lemma that if there exists a feasible algorithm for

for the construction of the functional $\varphi(i, y)$ satisfying the conditions of the lemma, then conditions 1 through 3 of the theorem are indeed the necessary conditions for an optimum. Such an algorithm, in particular, is defined by Bellman's method if the sets $V_y(i)$, $i=1, 2, \ldots, N$, coincide with Y and the functional $f^0(i, y, u)$ is bounded on $V(i)$ for every fixed $i \in A$.

Consider the auxiliary problem of minimizing the functional

$$\tilde{I} = \sum_{i=0}^{N} \tilde{f}^0(i, \ y(i), \ u\,(i))$$

on the set $\tilde{D}$ of the pairs $y(i)$, $u(i)$. Here $\tilde{f}^0[i, \ y, \ u] = f^0(i, \ y, \ u)$ for $y \in V_y(i)$ and $\tilde{f}^0 = K$ for $y \in V_y(i)$; $K = q + \sup\limits_{(y,u) \in V(i), i \in A} f^0(i, \ y, \ u)$; q is any number satisfying the inequality

$$q > \sup_{L} I - \inf_{E} I = \sum_{i=0}^{N} [\sup_{(y,u) \in V(i)} f^0(i, \ y, \ u) - \inf_{(y,u) \in V(i)} f^0(i, \ y, \ u)].$$

The set $\tilde{D}$ differs from D only in the structure of the set $\tilde{V}(i)$ of the admissible pairs $(y, u) \in Y \times U$. Indeed, $\tilde{V}_y(i)$ coincides with Y for all $i \in A$, and $\tilde{V}_u(i, y)$ coincides with $V_u(i, y)$ for $y \in V_y(i)$ and with U for $y \in Y/V_y(i)$. Clearly $D \subset \tilde{D}$ and $\tilde{I} = I$ for $(y(i), u(i)) \in D$. On any element from $\tilde{D}$ which does not belong to D, the functional $\tilde{I}$ assumes a value which is greater than its value on any element of D. Indeed, suppose that the condition $(y(i), u(i)) \in V(i)$ breaks down for $1 \leqslant m \leqslant N+1$ values of i. Then

$$\tilde{I} - \sup_{D} I \geqslant mq + \inf_{E} I - \sup_{E} I > (m-1)q \geqslant 0.$$

This signifies that starting with some $S = s$ all the elements of the minimizing sequence of the functional $\tilde{I}$ lie in D, i.e., $\{\bar{y}_S(i), \ \bar{u}_S(i)\} \subset D$ for $S > s$. Since $\tilde{I} = I$ on D, this sequence is also a minimizing sequence for the functional I.

Let the functional $\varphi(i, y)$ satisfy the conditions of the theorem for this problem. The functionals $R(i, y, u)$, Φ_0, and Φ_1 corresponding to this $\varphi(i, y)$ coincide on $V(i)$ with the analogous functionals of the original problem, and since $V(i) \subset \tilde{V}(i)$, they go on the sequence $\{\bar{y}_S(i), \ \bar{u}_S(i)\} \subset D$ to their largest (smallest) value on $V(i)$, i.e., the functional $\varphi(i, y)$ satisfies conditions 1 through 3 of Theorem 6.1 for the problem of minimizing I on D also. Since the problem of minimizing $\tilde{I}$ on $\tilde{D}$ belongs to the type of problems for which the existence of the functional $\varphi(i, y)$ is a necessary condition, it is also a necessary condition for the original problem.

Bibliography for Chapter VI

1. Bellman, R. Dynamic Programming. — Princeton Univ. Press. 1957.
2. Butkovskii, A. G. Teoriya optimal'nogo upravleniya sistemami s raspredelennymi parametrami (Theory of Optimal Control of Systems with Distributed Parameters). — Nauka. 1965.
3. Propoi, L. I. O printsipe maksimuma dlya diskretnykh sistem upravleniya (The Maximum Principle for Discrete Control Systems). — Avtomatika i Telemekhanika, 26, No. 7. 1965.

THE SIMPLEST FUNCTIONAL ON THE SET OF DIS-CONTINUOUS FUNCTIONS AND FUNCTIONS WITH A BOUNDED DERIVATIVE

The calculus of variations arose back in the 17th and the 18th century as a branch of mathematical analysis concerned with the search of objects (functions, curves, surfaces, etc.) on which a given integral attained its minimum or maximum value. Because of the then prevailing concepts in mathematics and the first particular applications of the theory (the brachistochrone problem, the problem of light refraction, the problem of isoperimeters, and later the least action principle), it was naturally assumed that the minimum (or the maximum) would be attained on a smooth or, in the worst case, piecewise-smooth continuous curve, provided the lower bound of the functional existed. This conviction persisted until the time of Weierstrass, and it actually served as a basis for Lagrange's method of variations, which reduced the problem of the extremum of a functional to a boundary-value problem for differential Euler—Lagrange equations whose solution was followed by a verification of a whole gamut of necessary and sufficient conditions. After Lagrange, considerable contributions to the theory were made by Legendre, Jacobi, Weierstrass, and Hilbert, who developed Lagrange's method to a stage when it became a highly refined, powerful, and often irreplaceable tool of analysis, which to this day constitutes the fundamental algorithm for the solution of variational problems.

However, Weierstrass in his famous disputes with Riemann produced numerous examples showing that the minimum of a functional $I(u)$, where u is an element of some set U, is not necessarily attained on the elements of that set. In particular, if $u \in C_1$, the minimizing solution does not always belong to the class C_1 of continuous smooth functions $y(t)$ and its derivative, and even the function itself, may have discontinuities. Moreover, as we shall see below, the function $y(t)$ may fail to represent a curve. Hence it follows that by confining the analysis to C_1 we cannot obtain a conclusive solution of the problem regarding the absolute minima of functionals.

This development was apparently responsible for Hilbert's unconventional approach to the problem of the absolute minimum of the functional

$$I = \int_A^B F(y, y', s)\, ds. \tag{S.1}$$

He rejected the entire refined, sophisticated, and rigorous classical
apparatus of the method of variations and defined the functional (S. 1) on the
set of all rectifiable curves. He then proceeded to construct these curves,
selecting those whose length — the functional I — approached a lower bound
(a minimizing sequence). As the next stage, he proved the existence of a
limit curve and established its extremal and functional properties. This
approach constituted the prototype of a new algorithm, one of the so-called
direct methods of variational calculus. We will concentrate on two trends
among the various ideas prevailing in this field.

    1. The theoretical-functional trend. This trend is mainly
concerned with the existence of an absolute extremal and the determination
of its functional properties. A characteristic feature of this trend is the
large-scale application of the theory of functions of a real variable. The
problem of minimum is considered on the set of absolutely continuous
functions, using Lebesgue integrals. This trend reached its peak in the
1920—1930's in the work of Tonelli and his school in the West and Krylov,
Bogolyubov, and Lavrent'ev in the USSR. Tonelli /17/ showed, for a
number of important particular cases, that the absolute extremum is
attained in the class of absolutely continuous functions. Lavrent'ev /10/
proved that the problem of the absolute minimum of the functional

$$I = \int_{a}^{b} F(t,\ y,\ y')\, dt \qquad\qquad (S.\ 2)$$

becomes meaningless if the class of admissible functions is extended to
include all functions of bounded variation. He also derived the sufficient
conditions for the lower limit of the functional (S. 2) in the class of absolutely
continuous functions to coincide with its lower limit in the class of
continuously differentiable functions. Later, this result was strengthened by
Tonelli. Under certain additional conditions imposed on $F(t,\ y,\ y')$, the
minimizing solution $y(t)$ has almost everywhere a first and a second derivative,
provided it belongs to the class of absolutely continuous functions.
Bogolyubov /1/ generalized Tonelli's results. We thus see that the
theoretical-functional school produced quite significant results. However,
the fundamental problem of this school — namely the problem of existence
of the absolute extremum and the functional properties of extremals —
remains on the whole unsolved even for the simplest functional (S. 2), not
to mention the practical construction of extremals. It is not clear under
what conditions the extremal belongs to the class of absolutely continuous
functions and when it does not belong to this class and, if the latter applies,
whether or not the class of the admissible objects can be extended so as
to include the objects on which the minimum is attained.

    The material of the following sections indicates that these problems
cannot be solved at all by traditional theoretical-functional methods. After
all, the choice of the class of admissible objects is prescribed not by the
inner logic of the variational calculus but by entirely extraneous factors
stemming from the elements of the theory of functions of a real variable.
Thus, the absolutely continuous functions are adopted as the class of admissible objects because the Lebesgue integral is defined on this class.
But this is clearly of no relevance for the fundamental processes of the
calculus of variations.

The above considerations reveal the limitations of the theoretical-functional methods, but in no way detract from their value, since in a number of important cases they provide the best and, possibly, the only tool for the solution of the problem of the existence and the properties of absolute extremals.

2. The second trend in the theory of direct methods can be described as the applied trend. The determination of a minimizing sequence directly involves the sought function on which the extremum is attained and calls for approximating this function by a suitable sequence. If a minimizing sequence can be effectively chosen, we automatically prove the existence and often approach (although purely theoretically) with any desired accuracy to the solution of the problem. Since the Euler—Lagrange equations are analytically solvable in exceptional cases only, this direct approximation is generally more effective than the solution of the corresponding boundary problem. The second trend is thus mainly concerned with the various aspects of the particular choice of minimizing sequences and their convergence.

Historically, the first method of construction of a minimizing sequence was developed by Euler (his finite-difference method). After two centuries in oblivion, it was resurrected and rigorously developed by Soviet mathematicians, mainly Lyusternik /11/, and also Petrovskii, Krylov, Bogolyubov. In the West, this method was taken up by Courant's school /8/.

A powerful direct method for the solution of variational problems is Ritz's method, first advanced in /16/. Ritz considers an n-parametric family of functions $y_n(t, a)$, $a = \{a_1, a_2, \ldots, a_n\}$. On these functions, the functional $I(y)$ is reduced to a function of a finite number of variables. The extremum $I(y_n) = I_n$ is found by determining the coefficients from the equations

$$\frac{\partial I_n}{\partial a_i} = 0, \quad i = 1, 2, \ldots, n.$$

In ordinary problems, the sequence of these extremal functions $y_n(t)$ was assumed to go in the limit to a function extremizing the functional $I(y)$. This method found wide applications in a variety of problems. Poincaré referred to it as a method for the engineer, thus stressing its outstanding applied importance.

Very extensive literature, both Soviet and Western, is currently available on Ritz's method. Of particular merit is the study of Krylov and Tamarkin /9/ who established a theoretical foundation for Ritz's method by proving the convergence of the minimizing sequence for a wide range of important applied problems.

The complete solution of the variational problem by the direct method is obtained by a combination of the two schools of thought and includes the following stages:

1. Construction of a minimizing sequence $\{y_n(t)\}$.

2. Proof of the convergence of $\{y_n(t)\}$ to a function $y(t)$ which belongs to the class of admissible curves.

3. Proof of the convergence

$$\inf_{y(t)} I(y(t)) = \lim_{n \to \infty} I(y_n(t)).$$

Significant difficulties are encountered already in the first stage, since we have to solve a system of n finite equations in n unknowns for every fixed $n \to \infty$. Since these equations, as a rule, are nonlinear, new solutions may arise at every successive stage, greatly complicating the situation. In general, a direct solution of the variational problem is more involved than the solution by the method of variations, since in addition to obtaining the extremal we at the same time solve the more general problem of its approximation by some given system of functions.

However, the solution of problems raised by Weierstrass's discovery does not necessarily follow Hilbert's path: the method of variations can be improved and generalized to a wider class of admissible curves. In this category we have the derivation of the well-known Erdmann—Weierstrass conditions at the corner points of the extremal and the theory of deeper discontinuities developed by the Soviet mathematician Razmadze /13/. Razmadze proceeded from Weierstrass's well-known example: the integral

$$I = \int_{-1}^{+1} t^2 y'^2 dt$$

with the boundary conditions $y(-1) = -1$, $y(1) = 1$ has a zero lower bound. Indeed, the value of this integral on the family of curves

$$y = \frac{\tan^{-1} \dfrac{t}{\varepsilon}}{\tan^{-1} \dfrac{1}{\varepsilon}}$$

goes to zero for $\varepsilon \to 0$, but the lower bound is not attained for any continuous curve. In other words, the problem has no classical solution. The limit of the family $y(t, \varepsilon)$ for $\varepsilon \to 0$ is the discontinuous function

$$\bar{y}(t) = \begin{cases} -1 & t < 0 \\ 1 & t > 0, \end{cases}$$

and $I(\bar{y}) = 0$. Thus, if the class of admissible functions is extended to include functions with a discontinuity of the first kind, the integral can be minimized. Razmadze raised a general question: in what cases a problem which is unsolvable in the class of continuous functions has a solution on a wider set of curves with one discontinuity point.

An integral of a discontinuous function is defined as a sum of the integrals over the continuous sections:

$$I(y) = \int_a^b F(t, y, y')dt = \int_a^{t_0 - 0} F(t, y, y')dt + \int_{t_0 + 0}^b F(t, y, y')dt, \qquad \text{(S. 3)}$$

where t_0 is a point of discontinuity of the first kind of $y(t)$. The functional of the discontinuous function $y(t)$ defined in this way satisfies the condition

$$I(y) = \lim_{n \to \infty} I(c_n), \qquad \text{(S. 4)}$$

where $\{c_n\} \subseteq C_1$ is an approximating sequence of continuous smooth functions, only if we have at the discontinuity point t_0

$$f(t_0,\ y,\ 0,\ 1) = \lim_{p \to 0} pF\left(t_0,\ y,\ \frac{1}{p}\right) = 0. \qquad (\text{S. }5)$$

The necessary conditions to be satisfied by the extremal at the discontinuity point are derived,

$$\left.\begin{aligned}
F_{y'}\big|_{t_0-0} = F_{y'}\big|_{t_0+0} = 0, \\
F\big|_{t_0-0} = F\big|_{t_0+0}.
\end{aligned}\right\} \qquad (\text{S. }6)$$

After that, the theory of sufficient conditions is developed.

In general, when condition (S. 4) does not restrict the class of admissible curves, the discontinuity is "floating", i. e., not fixed. In particular cases, when condition (S. 5) is satisfied only in isolated points of the segment $[a, b]$, the discontinuity point is fixed (Weierstrass's example belongs to this category).

Further development of the method of variations along the lines laid down by Razmadze was undertaken by Nikoladze /12/, Ermilin /2, 3, 4/, Kerimov /5, 6, 7/. Krylov generalized the fundamental lemma of variational calculus using Razmadze's results.

Razmadze's main achievement, in our opinion, is the inclusion of curves containing a finite number of vertical segments among the admissible curves. This generalization in itself does not resolve the difficulty, but it constitutes an important step forward toward the application of a new algorithm, described below, which ensures a complete solution of the problem of the absolute minimum for a wide class of functionals. This algorithm includes Razmadze's general case as a particular case and essentially advances the solution of the problem for functionals not included in this class. Particular results of Razmadze's method and the entire theory of necessary and sufficient conditions lose much of their value in this case, since both the absolute and the strong local minima are attained on the (y, z) minimals constructed by this algorithm and only on them. Razmadze's theory retains its original value only for the particular case of fixed discontinuity points.

Another, although less significant, point to be remembered is that definition (S. 3) of a functional on a discontinuous function is by no means unobjectionable. It suffices to note that this definition is meaningless even for the simplest problem of a curve of minimum length, since it ignores the length of the vertical sections.

A more general and more natural definition of a functional on discontinuous functions will be given below. It will include, as a particular case, those problems for which definition (S. 3) is meaningful.

A new method is proposed for the solution of variational problems. This method establishes the existence of a new class of minimizing solutions, which are no less typical then the classical Euler—Lagrange extremals, but are of fundamentally different nature. Unlike the Euler—Lagrange extremals, these new extremals are not solutions of any boundary value problem. Their finite equations are written directly in the form of the necessary conditions of extremum. The new minimizing solutions are not necessarily functions:

they may belong to an entirely new class of objects, called (y, z) curves.
If $z(t) \equiv y'(t)$ everywhere, the (y, z) curve is an ordinary piecewise-
smooth function. If, however, $y'(t) \neq z(t)$, the (y, z) minimal is not
a function, but if it is known, i. e., the pair of functions $y(t)$, $z(t)$ are known,
a minimizing sequence of piecewise-smooth curves approximating to the
(y, z) minimal is also known. The new method is considered for the particular case of the simplest functional

$$I = \int_a^b F(t, y, y')dt, \qquad (S. 7)$$
$$y(a) = a_1; \quad y(b) = b_1.$$

This functional is traditionally used as a touchstone for all new theories
in variational calculus (in many cases it is sufficient to reject a theory),
and more than half the publications on the subject are concerned with it.
If the function

$$f(t, y, p, 1) = pF\left(t, y, \frac{1}{p}\right)$$

for $p \neq 0$ exists and is continuous in p in the (t, y) plane, the proposed
algorithm gives a complete solution of the problem of the absolute minimum
of the functional (S. 7). In this case, the minimum is attained on the above-
mentioned minimals of the new type. An exceptionally simple necessary
and sufficient condition of minimum is derived in the form of a minimum
of some known function $S(t, y, z)$ with respect to y and z for every fixed
$t \in [a, b]$. In other cases, the minimum may be attained both on the (y, z)
minimals and on the classical Euler—Lagrange extremals.
Next we consider the important applied problem of the minimum of the
functional (S. 7) on the set of functions with a bounded derivative, i. e.,
functions satisfying inequality constraints. It is shown that in this case we
are again dealing with minimals which are analogs of the (y, z) minimals
of the simplest functional, but the role of the vertical directions is assumed
by the limit directions

$$y' = f_1(t, y) \quad \text{or} \quad y' = f_2(t, y).$$

Under certain conditions, the minimum in this problem is attained on
regular Euler extremals.
The new method is applied to solve some modern variational problems
of applied mechanics. These solutions are of independent interest and,
moreover, provide an excellent illustration of the application of the new
method and the actual form of the intangible (y, z) minimals.

Some introductory propositions

Let a functional I satisfying the condition

$$\inf_{u \in M} I(u) > -\infty \qquad (S. 8)$$

be defined on some set M.

We are looking for the absolute minimum of $I(u)$ over M, i. e., for an object $\bar{u}$ which satisfies the equality

$$I(\bar{u}) = \inf_{u \in M} I(u). \tag{S. 9}$$

The elements of the set M do not necessarily include the element $\bar{u}$ on which the lower bound is attained. In this case, we shall try to embed M in a wider set $\bar{M} \supset M$ in which the minimum element $\bar{u}$ is contained, appropriately extending the definition of the functional I on the set $\bar{M}$. The extension $I_{\bar{M}}$ (the definition of I on $\bar{M}$) only has to meet the following requirements:

1) on the elements of M it coincides with the original functional, i. e., $I_{\bar{M}}(u) = I_M(u)$, $u \in M$;

2) if $u \in \bar{M}$ there exists a sequence $\{u_n\} \subset M$ such that

$$I_{\bar{M}}(u) = \lim_{n \to \infty}(u_n). \tag{S. 10}$$

If a suitable set $\bar{M}$ is given (in the particular case $\bar{u} \in M$ it coincides with M), the element $\bar{u} \in \bar{M}$ minimizing the functional $I_{\bar{M}}(u)$ on $\bar{M}$ has been determined, and the sequence $\{\bar{u}_n\} \subset M$ satisfying the condition

$$I(\bar{u}) = \lim_{n \to \infty} I(\bar{u}_n) \tag{S. 11}$$

has been constructed, we consider the problem of the absolute minimum of the functional on the set M solved.

The element $\bar{u} \in \bar{M}$ is called the absolute minimizing solution (or the absolute minimal) of the functional I on M. The sequence $\{\bar{u}_n\}$ is a minimizing sequence, and the set $\bar{M}$ is an I-extension of the set M.

L e m m a . Let $\bar{u} \in \bar{M}$ satisfy (S.9). Then

$$I(\bar{u}) = \inf_{u \in \bar{M}} I(u). \tag{S. 12}$$

Conversely, if $\bar{u} \in \bar{M}$ satisfies (S. 12), (S. 9) holds true. Let (S. 9) be true, and suppose that (S. 12) is not satisfied. Then there exists an element $v \in \bar{M}$ such that $I(v) < I(\bar{u})$. We write

$$I(\bar{u}) - I(v) = p > 0.$$

In virtue of the definition of I on $\bar{M}$, there exists a sequence $\{u_n\} \subset M$ such that $|I(v) - I(u_n)| < P$ for sufficiently large n and therefore $I(u_n) < I(\bar{u})$. The last inequality contradicts (S. 9), however. Let (S. 12) be true. Since $M \subset \bar{M}$, we have $I(\bar{u}) \leqslant I(u)$, $u \in M$. On the other hand, according to the definition of I on $\bar{M}$, there exists a sequence $\{u_n\} \subset M$ such that $I(u_n) = I(\bar{u})$. Hence, in virtue of the definition of the lower bound, we obtain (S. 9).

This lemma shows that the minimals of the functional I on the sets M and $\bar{M}$ coincide, so that instead of minimizing the functional on the set M we may minimize I on $\bar{M}$, if this presents any advantages.

§ S. 1.   THE SIMPLEST FUNCTIONAL

1. 1.   Statement of the problem

Consider the minimum of the functional

$$I(u) = \int_a^b F(t, y, y')\, dt \qquad\qquad \text{(S. 13)}$$

on the set U of curves with the following properties:
1) the coordinates t and y of the points of the curve $u \in U$ may be represented as continuous functions of some parameter k ;
2) the function $y(t)$ is continuous along the curve u and is single-valued everywhere on $[a, b]$, except a finite set of points $\{\mu_i\}$ $(i=1, 2, \ldots, k)$, where it may have discontinuities of the first kind;
3) the derivative $y'(t)$ of the function $y(t)$ is continuous and bounded on the set $[a, b]/\{\mu_i\}$;
4) the function $y(t)$ satisfies the condition

$$y(a) = a_1;\ y(b) = b_1, \qquad\qquad \text{(S. 14)}$$

where a_1 and b_1 are known;
5) the curves $u \in U$ lie in the region B on the (t, y) plane where the function $F(t, y, z)$ is continuous in all the three arguments together with the derivatives F_t, F_y, F_z for any z;
6) the upper and the lower boundaries $\Gamma_1(t)$ and $\Gamma_2(t)$ of the region B, if they exist, have the properties $1-3$ of the set U.
Definition 1.   The right-hand side of (S. 13) is interpreted as a line integral along the curve u in the direction from the point $A(a, a_1)$ to the point $B(b, b_1)$, i. e.,

$$I(u) = \int_\alpha^\beta f(t, y, \dot{t}, \dot{y})\, dk, \qquad\qquad \text{(S. 15)}$$

where

$$f(t, y, p, q) = pF\left(t, y, \frac{q}{p}\right); \qquad\qquad \text{(S. 16)}$$

k is the parameter taking the values $k=\alpha$ for $t=a$, $y=a_1$ and $k=\beta>\alpha$ for $t=b$, $y=b_1$ which increases along u from A to B.
In virtue of the above properties of the function $F(t, y, z)$ and the set u, the latter contains curves on which the functional is finite. For the problem to be meaningful, we have to assume further that

$$\inf_{u \in U} I(u) = m > -\infty. \qquad\qquad \text{(S. 17)}$$

Taking $dk = dt$ for $\mu_i < t < \mu_{i+1}$ and $dk = |dy|$ for $t = \mu_i$, $i = 0, 1, \ldots, k$, where $\mu_0 = a$, $\mu_k = b$, we obtain from Definition 1

$$I(u) = \sum_{i=0}^{n-1} I_i^L + \sum_{i=0}^{n} \Phi\left[\mu_i,\, y(\mu_i),\, \overline{y}(\mu_i)\right], \qquad (\text{S. }18)$$

where

$$I_i^L = \int_{\mu_i+0}^{\mu_{i+1}-0} F\,dt; \qquad (\text{S. }19)$$

$$\Phi(t,\, y,\, \overline{y}) = \int_{\overline{y}}^{y} W(t,\, \xi,\, \operatorname{sign}(y-\overline{y}))\,d\xi; \qquad (\text{S. }20)$$

$$y = y(t+0); \quad \overline{y} = y(t-0); \qquad (\text{S. }21)$$

$$W(t,\, y,\, \operatorname{sign}(y-\overline{y})) = \lim f(t,\, y,\, p,\, 1);$$

$$|p| \to 0, \quad \operatorname{sign} p = \operatorname{sign}(y-\overline{y}). \qquad (\text{S. }22)$$

Because of the properties of the function $F(t,\, y,\, z)$ and the set U, the integrals I_i^L always exist. Therefore $I(u)$ exists on those and only those curves $u \in U$ for which the integrals (S. 20) exist on the vertical sections (if any).

Let the curve $\overline{u} \in U$ be the minimal of $I(u)$ in U, i. e., $I(\overline{u}) = \inf_{u \in U} I(u)$. By (S. 17), $I(\overline{u})$ exists and the existence of the integrals $\Phi(\mu_i,\, y_i,\, \overline{y}_i)$ on the vertical sections therefore may be regarded as the first necessary condition to be satisfied by the minimizing solution $\overline{u}$.

This one condition enables us, in some highly important particular cases, to assess the qualitative character of the minimum and, in particular, the possible existence of discontinuities and their position.

The general case in our treatment is such that the condition of existence of $\Phi(t,\, y,\, \overline{y})$ at the points of discontinuity does not impose any additional constraints on the class of admissible curves, i. e., $\Phi(t,\, y,\, \overline{y})$ exists for all $t,\, y,\, \overline{y} \in B$.

1. 2. Relation to the problem of the absolute minimum in the class of piecewise-smooth functions C_1

In the class of piecewise-smooth functions C_1, Definition 1 coincides with the conventional classical Riemann integral $I(u),\, u \in C_1$.

Let the object $\overline{u} \in \overline{U}$ be the absolute minimal of I on $\overline{U}$, where $\overline{U}$ is some I-extension of U. The set $\overline{U}$ is the I-extension of the class C_1 if for every curve $u \in U$ there exists a sequence $\{c_n\} \subset C_1$ such that

$$I_u(u) = \lim_{n \to \infty} I(c_n) \qquad (\text{S. }23)$$

Here the subscript u indicates that the functional is taken in the sense of (S. 15). Indeed, in virtue of the definition of the I-extension, for any $u \in U$ there exists a sequence $\{u_n\} \subset U$ satisfying (S. 12) and, therefore, by (S. 23), also a sequence $\{c_n\} \subset C_1$ such that

$$I_{\overline{u}}(u) = \lim_{n \to \infty} I(c_n^-).$$

By our lemma, the absolute minimal $\bar{u}$ of the functional I in u is therefore also the absolute minimal in the class C_1, i. e.,

$$I\left(\bar{u}\right)=\inf_{u \in C_1} I\left(u\right), \tag{S.24}$$

if for every $u \in U$ there exists a sequence $\{c_n\} \subset C_1$ satisfying (S. 23).

We will show in what follows that this condition is always satisfied in the general case. In the so-called particular cases, it is not always satisfied.

If the absolute minimal in U is also the absolute minimal in C_1, we will refer to it as the absolute minimal of the functional I (without mentioning any classes).

1. 3. The general case

The general case in our terminology is such that the condition of existence of $\Phi(t, y, \bar{y})$ on the vertical sections does not impose any additional restrictions on the class of curves, i. e., $\Phi(t, y, \bar{y})$ exists for all $t, y, \bar{y}$. This implies that for every fixed $t \in [a, b]$, the two functions of a single variable $f[t, y, \pm 0,1]$ should be summable on any segment $[y, \bar{y}] \in [\Gamma_2(t), \Gamma_1(t)]$. In what follows, we assume that both functions $f(t, y, \pm 0,1)$ moreover exists and are continuous with their derivatives f_t and f_y everywhere in B.

We will see below that the character of the extremals is determined by the properties of the function $f(t, y, p, 1)$ for $p=0$, specifically, by whether the function $f(t, y, p, 1)$ for $p=0$ is continuous or discontinuous in p. Both these cases are of the greatest significance both from theoretical and applied considerations, and we will consider them separately in some detail.

The system of definitions and theorems introduced below leads to a complete solution of the problem of absolute minimum of the functional (S. 13) in the first case and discloses a number of highly valuable new facts in the second. It admits of very extensive generalizations and provides a foundation for all further constructions (Figure S. 1).

Consider the set of polygonal lines , $\{\gamma_n\} \subset U_1$, described by the functions $y_n(t)=y_i+y_i'(t-t_i)$ for $t_i \leqslant t < t_{i+1}$, $i=0, 1,\ldots, n-1$, where

$$a=t_0 < t_1 < \ldots < t_{n-1} < t_n = b$$

are the abscissas of the discontinuity points of $y_n(t)$, and y_i, y_i' are $2n$ independent parameters defining the polygonal line γ_n for any fixed partition.

Definition 2. Consider a line $u \in U$ defined by the function $y(t)$. We shall say that the sequence $\{\gamma_n\}$ approximates the line u, or $\{\gamma_n\} \underset{n \to \infty}{\longrightarrow} u$, if

$$y_i = y(t_i+0), \tag{S.25}$$

and for any $\varepsilon > 0$,

$$|t_{i+1}-t_i| < \varepsilon; \quad \left|y_i' - y'(t_i+0)\right| < \varepsilon \tag{S.26}$$

for $n>N(\varepsilon)$.   Here $i=0, 1, \ldots, n-1$.   In this sense, $\{\gamma_n\}$ is everywhere dense on U.

Definition 3.   Consider a pair of functions $y(t) \in U$ and $z(t)$ which are defined, bounded, and continuous on the set $[a, b]/\{\mu_i\}$.   We construct the sequence $\{\gamma_n\}$ so that

$$y_i = y(t_i+0) \tag{S. 27}$$

and for any $\varepsilon>0$

$$|t_{i+1}-t_i| < \varepsilon; \; |y_i' - z(t_i+0)| < \varepsilon \tag{S. 28}$$

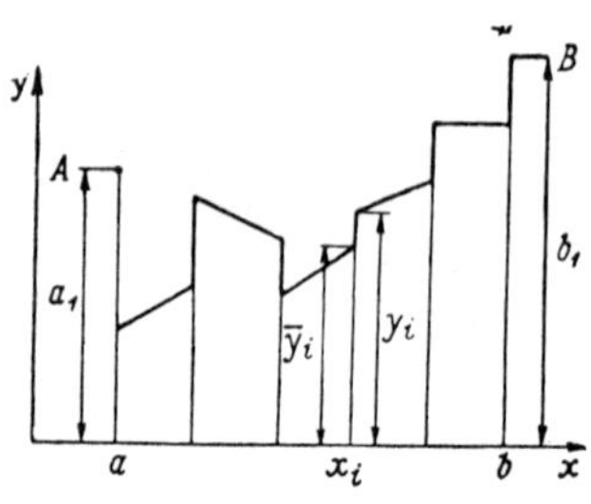

FIGURE S. 1

for $n>N(\varepsilon)$, $i=0, 1, \ldots, n-1$.   In this case we say that the sequence $\{\gamma_n\}$ approximates the (y, z) line u_0, or $\{\gamma_n\} \to u_0$.   The set of these (y, z) lines is designated U_0.   The function $y(t)$ is the zero closeness function of the (y, z) line u_0 and $z(t)$ is the local slope of this line.

If

$$z(t) = y'(t) \tag{S. 29}$$

almost everywhere on $[a, b]$, we have $u_0 \in U$.   Thus $U \subset U_0$.

Definition 4.   The functional (S. 13) of a (y, z) line is defined as the limit

$$I(u_0) = \lim_{\{\gamma_n\} \to u_0} I(\gamma_n). \tag{S. 30}$$

Theorem 1.   If the function $f(t, y, p, 1)$ exists and is continuous in p for $p=0$ and for any $t, y \in B$, the functional $I(u_0)$, $u_0 \in U_0$, exists and is expressed by the formula

$$I(u_0) = \sum_{i=0}^{k-1} \int_{\mu_i+0}^{\mu_{i+1}-0} [F(t, y, z) + W(t, y)(y'-z)]\, dt + \sum_{i=0}^{k} \Phi(\mu_i, y_i, \bar{y}_i) \tag{S. 31}$$

or by

$$I(u_0) = (R)\int_a^b S(t, y, z)\, dt + \Phi(b, b_1, c(b_1)) - \Phi(a, a_1, c(a_1)), \tag{S. 32}$$

where

$$S(t,\ y,\ z) = F(t,\ y,\ z) - W(t,\ y)(z - c'(t)) - \int_{c(t)}^{y} W_t(t,\ \xi)\, d\xi, \qquad (\text{S. }33)$$

where $c(t) \in C_1$ is an arbitrary function; $W(t,\ y)$ and $\Phi(t,\ y,\ c)$ are defined by (S. 20) and (S. 22); μ_i are the discontinuity points of $y(t)$, $i = 0,\ 1,\ \ldots,\ k$.

The symbol (R) indicates that a Riemann integral is meant. Since $f(t,\ y,\ p,\ 1)$ is continuous for $p = 0$,

$$W(t,\ y,\ 1) = W(t,\ y,\ -1) = W(t,\ y) = f(t,\ y,\ 0,\ 1). \qquad (\text{S. }34)$$

Using (S. 20) and (S. 34), we may write

$$\Phi(t,\ y,\ \bar{y}) = \Phi(t,\ y,\ c) - \Phi(t,\ \bar{y},\ c), \qquad (\text{S. }35)$$

where $c(t)$ is an arbitrary function. In what follows we take $c(t) \in C_1$.

By (S. 18) and (S. 35),

$$I(\gamma_n) = \sum_{i=0}^{n-1} S_i \Delta t_i + \Phi(b,\ b_1,\ c(b)) - \Phi(a,\ a_1, c(a)), \qquad (\text{S. }36)$$

where

$$S_i \Delta t_i = \int_{t_i+0}^{t_{i+1}-0} F(t,\ y_n,\ y_n')\, dt + \Phi(t_i,\ y_i,\ c) -$$
$$- \Phi(t_{i+1},\ \bar{y}_{i+1},\ c);$$

t_i are the discontinuity points of the function $y_n(t)$, $i = 0,\ 1,\ \ldots,\ n-1$, or using Lagrange's theorem of finite differences

$$S_i = F[t_i^*,\ y_n(t_i^*),\ y_n'(t_i^*)] - \Phi'(t_i^*,\ y_n(t_i^*),\ c(t_i^*)) =$$
$$= F[t_i^*,\ y_n(t_i^*),\ y_n'(t_i^*)] - W(t_i^*,\ y(t_i^*)[y_n'(t_i^*) -$$
$$- c'(t_i^*)] - \int_{c(t^*)}^{y_n(t^*)} W_t(t^*,\ \xi)\, d\xi; \qquad (\text{S.}37)$$
$$t_i^* = t_i + \Delta t_i \theta,\ 0 < \theta < 1.$$

Using (S. 27) and (S. 28), we may write

$$I(u_n) = \sum_{i=0}^{n-1} S[t^*,\ y(t^*),\ z(t^*)] \Delta t_i + 0(\varepsilon) +$$
$$+ \Phi(b,\ b_1,\ c(b)) - \Phi(a,\ a_1,\ c(a)). \qquad (\text{S. }38)$$

Taking the limit $\gamma_n \to u_0$ in (S. 30), some terms in (S. 38) go to infinity, and $|t_{i+1} - t_i| \to 0$.

In virtue of the properties of $F(t,\ y,\ z)$, the function $S[t,\ y(t),\ z(t)]$ is continuous in all its three arguments in region B of the $(t,\ y)$ plane for any finite z. Since $y(t)$ and $z(t)$ are continuous almost everywhere and bounded on $[a,\ b]$, $S[t,\ y(t),\ z(t)]$ is also continuous almost everywhere, so that by the

Lebesgue theorem, the function $S(t, y(t), z(t))$ is Riemann-integrable. This indicates that for $\{\gamma_n\} \to u_0$ the sum (S. 38) has a limit, i. e., it is independent of the particular choice of the sequence $\{\gamma_n\} \to u_0$ (independent of the partition $\{t_i^*\}$ and the position of the points t_i^*).

We proved the existence of the functional $I(u_0)$ on any (y, z) line $u_0 \in U_0$ and the validity of (S. 31). Similarly, starting with (S. 18), we prove the validity of (S. 31) (this formula is proved in a more general form in Theorem 2). Q. E. D.

T h e o r e m 2. If the function $f(t, y, p, 1)$ has a discontinuity of the first kind in p for $p=0$ for any $(t, y) \in B$, the functional $I(u_0)$, $u_0 \in U_0$, exists and is expressed by the equality

$$I(u_0) = \sum_{i=0}^{k-1} \int_{\mu_i+0}^{\mu_{i+1}-0} [F(t, y, z) + W(t, y, \operatorname{sign}(y'-z))] \times$$

$$\times (y'-z)\,dt + \sum_{i=0}^{k} \int_{\underline{y}_i}^{\overline{y}_i} W(\mu_i \xi, \operatorname{sign}(y_i - \overline{y}_i))\,d\xi \tag{S. 39}$$

or by the equality

$$I(u_0) = (R) \int_a^b S(t, y, z, \operatorname{sign}(y'-z))\,dt +$$

$$+ \sum_{j=1}^{r} [\Phi(t_j, y, c, \operatorname{sign}(y'-\overline{z})) - \Phi(t_j, y, c, \operatorname{sign}(y'-z))] +$$

$$+ \Phi[b, b_1, c, \operatorname{sign}(b_1 - \overline{y}(b))] - \Phi[a, a_1, c, \operatorname{sign}(y(a) - a_1)], \tag{S. 40}$$

t_j are the points where the difference $y'-z$ reverses its sign, $j = 1, 2, \dots, r$.

By (S. 18), (S. 27), and (S. 28), remembering that $y(t)$, $z(t)$ are continuous and bounded, we have

$$I(\gamma_n) = \sum_{i=0}^{n-1} \int_{t_i+0}^{t_{i+1}-0} F(t, y_n, z)\,dt + \sum_{i=0}^{n} \int_{\underline{y}_i}^{\overline{y}_i} W[t_i, y, \operatorname{sign}(y_i - \overline{y}_i)]\,dy =$$

$$= \sum_{i=0}^{n-1} \{F(t_i, y(t_i), z(t_i)) +$$

$$+ W[t_i, y(t_i), \operatorname{sign}(y'(t_i) - z(t_i) + 0(\varepsilon))] + 0(\varepsilon)\} (t_{i+1} - t_i) +$$

$$+ \sum_{i=0}^{n} \int_{\underline{y}(\mu_i)}^{y(\mu_i)} W[\mu_i, \xi, \operatorname{sign}(y_i - \overline{y}_i)]\,d\xi + 0(\varepsilon). \tag{S. 41}$$

In virtue of the properties of the functions $F(t, y, z)$, $W(t, y, \pm 1)$, $y(t)$, and $z(t)$, the function

$$F(t, y(t), z(t)) + W[t, y(t), \operatorname{sign}(y'(t) - z(t))](y'(t) - z(t))$$

is bounded and continuous almost everywhere in the intervals (μ_i, μ_{i+1}). Hence, by the Lebesgue theorem, it is Riemann-integrable, i. e., the limit (S. 30), where $I(\gamma_n)$ is expressed in the form (S. 41), exists and is independent of the choice of the sequence $\gamma_n \to u_0$.

The validity of (S. 40) is proved along the same lines as Theorem 1.

Corollary 1. The functional $I(u_0)$, $u_0 \in U_0$, depends on two independent functions $y(t)$ and $z(t)$ whose derivatives either do not enter the integrand, as in (S. 31), or only enter as sign $(y'-z)$, as in (S. 40).

Corollary 2. It follows directly from expressions (S. 31) and (S. 39) for $I(u_0)$ that the functional is continuous on the set U_0 of (y, z) lines in the sense that

$$\left| I(\bar{u}_0) - I(u_0) \right| < \varepsilon, \left.\vphantom{\begin{array}{c}a\\b\end{array}}\right\}$$

if

$$\left| \bar{y}(t) - y(t) \right| < \eta(\varepsilon); \quad \left| \bar{z}(t) - z(t) \right| < \eta(\varepsilon) \tag{S. 42}$$

everywhere on $[a, b]$, except the η-neighborhoods of the points μ_i where $y(t)$ is discontinuous.

Corollary 3. It follows from (S. 32) and (S. 40) that Definitions 1 and 4 coincide on the set of $y(t)$.

Corollary 4. It follows directly from Definition 4 and Corollary 3 that the set U_0 is an I-extension of the set U.

Remark. Expression (S. 40) for $I(u)$, $u \in U_0$, is not defined on ordinary smooth curves, i.e., for $y'(t) = z(t)$. The continuity of the functional on U_0 shows that its definition can be extended for these curves if we take for $y' - z = 0$

$$\text{sign}(y'-z) = 1, \quad \text{or} \quad \text{sign}(y'-z) = -1.$$

In this case, expression (S. 40), like (S. 39), is defined over the entire set U_0.

We will now show that U, and also U_0, are I-extensions of the class C_1 of piecewise-smooth functions, i.e., for any line $u \in U$ there is a sequence $\{c_n\} \in C_1$ such that

$$I_u(u) = \lim_{n \to \infty} I(c_n). \tag{S. 23}$$

Consider some line $u \in U$. We replace all the vertical segments of the line u to segments making an angle of $\dfrac{1}{n}$ to the vertical which pass through the midpoints of the vertical segments. Let $n > 0$ if the deflection from the vertical is in the clockwise direction, and $n < 0$ if the deflection is counterclockwise. The deflection should be such that the resulting line c_n corresponds to a single-valued function $y_n(t)$, i.e.,

$$\text{sign}\, h = \text{sign}\, [y(\mu_i + 0) - y(\mu_i - 0)]. \tag{S. 42a}$$

Clearly $\{c_n\} \subset C_1$; then

$$
\left.
\begin{aligned}
I(c_n) &= \sum_{i=0}^{k-1} \int_{\mu_i+0\left(\frac{1}{n}\right)}^{\mu_{i+1}-0\left(\frac{1}{n}\right)} F\,dt \; + \\
&+ \sum_{i=0}^{k} \int_{\mu_i-0\left(\frac{1}{n}\right)}^{\mu_i+0\left(\frac{1}{n}\right)} F(t,\,y,\,n)\,dt = \sum_{i=0}^{k-1} \int_{\mu_i+0\left(\frac{1}{n}\right)}^{\mu_{i+1}-0\left(\frac{1}{n}\right)} F\,dt + \\
&+ \sum_{i=0}^{k} \int_{y\left(\mu_i-0\left(\frac{1}{n}\right)\right)}^{y\left(\mu_i+0\left(\frac{1}{n}\right)\right)} F(t,\,y,\,n)\,\frac{1}{n}\,dy; \\
&\qquad \operatorname{sign} n = \operatorname{sign} dy.
\end{aligned}
\right\}
\qquad (S.\ 43)
$$

Taking the limit $n \to \infty$ in (S. 43), we obtain on the right expression (S. 18) for $I(u)$ in the sense of Definition 1. Hence, for any line $u \in U$, there exists a sequence $\{c_n\} \subset C_1$ which satisfies (S. 23). Therefore, the absolute minimal on U is at the same time the absolute minimal of (S. 13) on C_1.

In our definition of U_0, we imposed the following conditions on its elements, $(y,\,z)$ lines: $y(t) \in U$ and $z(t)$ is bounded everywhere on $[a,\,b]$ and continuous on $[a,\,b]/\{\mu_i\}$, where $\{\mu_i\}$ is a finite set of points.

Definition 5. We say that the sequence of $(y,\,z)$ lines $\{u_n\} \in U_0$ goes to a $(y,\,z)$ object $u \in \bar{U}_0$ if for fixed t, almost everywhere on $[a,\,b]$,

$$
\left.
\begin{aligned}
y(t) &= \lim_{n \to \infty} y_n(t); \\
z(t) &= \lim_{n \to \infty} z_n(t).
\end{aligned}
\right\}
\qquad (S.\ 44)
$$

The last equalities are to be interpreted in the following sense: if $y(t)$ is finite at a point t, then for any given $\varepsilon > 0$ there exists N such that

$$
|y_n(t) - y(t)| < \varepsilon \quad \text{for} \quad n > N. \qquad (S.\ 45)
$$

If $y(t) = \pm\infty$, then for any $M > 0$ or, respectively, $M < 0$, there exists N such that for $n > N$

$$
y_n(t) > M > 0, \qquad (S.\ 46)
$$

or, respectively,

$$
y_n(t) < M < 0. \qquad (S.\ 47)
$$

The convergence for $z_n(t)$ is similarly interpreted.

The set $\bar{U}_0$ of $(y,\,z)$ objects is called the closure of U_0. Evidently, $\bar{U}_0$ is an I-extension of U_0.

D e f i n i t i o n   6. The functional (S.13) on an object $u \in \bar{U}_0$ is defined by the equality

$$I(u) = \lim_{\{u_n\} \to u,\, \{u_n\} \subset U_0} I(u_n). \tag{S. 48}$$

Here $I(u_n)$ is expressed by any one of the equalities (S. 31), (S. 32), (S. 39), (S. 40).

Let us now consider separately the two cases when $f(t, y, p, 1)$ is continuous at $p=0$ and when it has a discontinuity at $p=0$.

1. The function $f(t, y, p, 1)$ is continuous in p for $p=0$ and for any $(t, y) \in B$.

The problem of the absolute minimum of the functional (S. 13) in this case is completely solved by the following theorem.

T h e o r e m   3. The absolute minimum of the functional (S. 13) is attained on a (y, z) object $\bar{u} \in \bar{U}_0$ which satisfies the condition

$$S(t, \bar{y}, \bar{z}) = \inf_{\Gamma_2(t) \leqslant y(t) \leqslant \Gamma_1(t);\ -\infty \leqslant z \leqslant \infty} S(t, y, z) \tag{S. 49}$$

for fixed t almost everywhere on $[a, b]$ and only on this object.

P r o o f. We will first prove the theorem for the absolute maximum on U_0. By (S. 17) and the lemma we have

$$\inf_{u \in U_0} I(u) = m > -\infty. \tag{S. 50}$$

According to the definition of a lower bound, there exists a sequence $\{u_n\} \subset U_0$ such that $I(u_n) \underset{n \to \infty}{\longrightarrow} m$, $I(u_n) \leqslant m$. Expression (S. 31) for $I(u)$, $u \in U_0$, and (S. 50) directly show that:

1. The lower bound of $S(t, y, z)$ exists everywhere on $[a, b]$,

$$l(t) = \inf_{\Gamma_1(t) \leqslant y \leqslant \Gamma_2(t),\ -\infty \leqslant z < +\infty} S(t, y, z)$$

and since $S(t, y, z)$ is continuous in y and z, there exist $\bar{y}(t)$ and $\bar{z}(t)$ on which $S = l(t)$.

2. On the sequence $\{u_n\}$ for $n \to \infty$,

$$S(t, y_n, z_n) \to l(t) \quad \text{and} \quad y_n(t) \to \bar{y}(t),\ z_n(t) \to \bar{z}(t)$$

in the sence of Definition 5. Since $(y_n(t), z_n(t)) \in U_0$, the (y, z) object $\bar{u}$ described by the pair $(\bar{y}(t), \bar{z}(t))$ belongs to $\bar{U}_0$.

3. From the definition of the lower bound and from item 2 above we have

$$m = \lim_{n \to \infty,\, \{u_n\} \in U_0} I(u_n) = \lim_{y_n(t) \to \bar{y}(t);\, z_n(t) \to \bar{z}(t)} \int_a^b S(t, y_n, z_n)\, dt + \text{const.} \tag{S. 51}$$

By Definition 6, the last expression is the functional $I(u)$, $\bar{u} \in \bar{U}_0$.

We proved that the absolute minimum of the functional (S. 13) on $\bar{U}_0$ is attained on the (y, z) object $\bar{u} \in \bar{U}_0$ satisfying (S. 49) and, by item 2, only on this object. The theorem is thus valid for minima on U_0. Since U_0, and hence $\bar{U}_0$, are I-extensions of U and C_1, our lemma shows that the theorem remains true for the absolute minimum also. Q. E. D.

Corollary 1. It follows from (S. 49) and the continuity of $S(t,\ y,\ z)$ and its partial derivatives, that if the zero closeness function $\bar{y}(t)$ of the minimal $\bar{u}$ lies inside the region B, and the local slope $\bar{z}(t)$ is finite, then almost everywhere on $[a,\ b]$

$$S_y \equiv F_y - W_y \cdot z - W_t = 0; \quad \left.\begin{matrix} \\ \\ \end{matrix}\right\} \qquad\qquad \text{(S. 52)}$$
$$S_z \equiv F_z - W = 0.$$

Corollary 2. If the classical Euler–Lagrange extremal $\bar{y}(t)$ which is the solution of the equation

$$\delta I(y(t)) = 0$$

in the class C_1 of continuous piecewise-smooth functions does not satisfy (S. 49) ($\bar{z} = \bar{y}'$ in this case), the functional I either has a weak local minimum in C_1 on this extremal or has no minimum altogether. No strong local minimum in C_1, and certainly no absolute minimum, can be attained on this extremal.

Discussion. Condition (S. 49) defines a $(y,\ z)$ object $\bar{u}$ on which the functional (S. 13) attains its absolute minimum, i. e., it defines the zero closeness function $\bar{y}(t)$ of the functional and the local slope $\bar{z}(t)$. If we find that $\bar{y}(t) \in U$ and $\bar{y}'(t) = \bar{z}(t)$ everywhere on $[a,\ b]$, except a finite number of points, then $\bar{u} \in U$. Otherwise, the minimal $\bar{u} \bar{\in} U$, but the solution $\bar{y}(t)$, $\bar{z}(t)$ defines a minimizing sequence $\{\bar{\gamma}_n\} \in U$. If $\bar{u} \in U_0$, the minimizing sequence $\{\gamma_n\} \subset U$ is constructed as indicated in Definition 2, i. e., we construct the polygonal lines γ_n so that

$$y_i = \bar{y}(t_i); \quad y_i' = \bar{z}(t_i),\ i = 0,\ 1,\ 2, \ldots,\ n-1.$$

Irrespective of the partition of the segment $[a,\ b]$,

$$I(\gamma_n) \underset{n \to \infty}{\to} I(\bar{u}).$$

If $\bar{u} \in U_0$, we first select a sequence of pairs $\{y_m(t),\ z_m(t)\} \subset U_0$ which goes to $\bar{u}$ in the sense of Definition 5, and on each element of the sequence we construct a sequence of polygonal lines $\{\gamma_{nm}\} \to u_m \in U_0$ in the sence of Definition 3. Any sequence $\{\gamma^k\}$ selected so that $m \to \infty,\ n \to \infty$ for $k \to \infty$ is a minimizing sequence.

Thus, if we found the extremal pair $\bar{y}(t)$, $\bar{z}(t)$ we have all the elements of the minimizing sequence.

If the minimal $\bar{u} \bar{\in} U$, the minimizing sequence is of special importance, since the minimal itself cannot be constructed.

Let us investigate the minimals in more detail, assuming additionally that $F(t,\ y,\ z)$ and $W(t,\ y)$ are twice differentiable in all their three arguments in region B of the $(t,\ y)$ plane for any z.

1. Let (S.52) have a unique solution $y^0(t)$, $z^0(t)$ and a positive determinant

$$D(t, \ y^0, \ z^0) \equiv S_{yy} \cdot S_{zz} - S_{yz}^2 > 0. \tag{S. 53}$$

Since the right-hand sides of (S. 52) are continuous and have continuous partial derivatives, the existence theorem of implicit functions indicates that $y^0(t)$, $z^0(t)$ are continuous and differentiable on $[a, b]$. If $\Gamma_2(t) < y^0(t) < \Gamma_1(t)$ everywhere on $[a, b]$, $y^0(t)$ coincides with the zero closeness function $\bar{y}(t)$ of the minimal and $z^0(t)$ coincides with its local slope $\bar{z}(t)$ everywhere on $[a, b]$. At the points $t=a$ and $t=b$, the values of $\bar{y}$ and y^0 in general do not coincide:

$$\bar{y}(a) = a_1 \neq y_0(a).$$

Thus, if $D(t, \ y^0, \ z^0) > 0$, the minimal $\bar{u}_0$ not only belongs to U_0, but its zero closeness function $\bar{y}(t)$ and the local slope $\bar{z}(t)$ are continuous and differentiable everywhere on $[a, b]$, except the points $t=a$ and $t=b$, where $\bar{y}(t)$, in general, may have discontinuities of the first kind. If a posteriori we find that $\bar{y}'(t) = \bar{z}(t)$ almost everywhere on $[a, b]$, then $\bar{u} \in U$.

2. Now suppose that system (S. 52) has a finite number of solutions for which $D \neq 0$ almost everywhere on $[a, b]$. By the existence theorem of implicit functions, these solutions are piecewise-differentiable. The minimal $\bar{y}(t)$, $\bar{z}(t)$ consists of the sections of these solutions with $D > 0$ and of sections of the boundary. The selection of these sections is determined by the sufficient and the necessary condition (S. 49). It follows from the same condition that, besides the end points, the functions $\bar{y}(t)$ and $\bar{z}(t)$ may have discontinuities of the first kind at points $t = \mu_i$, where

$$S(\mu_i, \ y^{(1)}(\mu_i), \ z^{(1)}(\mu_i)) = S(\mu_i, \ y^{(2)}(\mu_i), \ z^{(2)}(\mu_i)). \tag{S. 54}$$

Here $y^{(1)}(t)$, $z^{(1)}(t)$ and $y^{(2)}(t)$, $z^{(2)}(t)$ are different solutions of (S. 52) or pieces of the boundary. At those points where (S. 54) is not satisfied but $D=0$, $\bar{y}'(t)$ or $\bar{z}'(t)$ may become infinite or not exist at all. Moreover, at some points of the segment $[a, b]$ or on some continuum, we may find that condition (S.49) is not satisfied for any of the solutions of (S.52), and inf $S(t, y, z)$ is attained for $z = \pm\infty$ or, if B is unbounded, for $y = \pm\infty$ (S remaining bounded). It follows that in this case the minimal $\bar{u}$ does not necessarily belong either to U_0 or to $\overline{U}_0$.

By Theorem 3, it always belongs to $\overline{U}_0$. As in previous cases, the minimal $\bar{z}(t)$, $\bar{y}(t)$ is fully defined by (S. 49).

Together with region B, we introduce a right cylinder Q in the (t, y, z) space having B as its base in the (t, y) plane. According to the statement of our problem, the function $F(t, y, z)$ is continuous in Q together with its derivatives F_t, F_y, F_z.

3. System (S. 52) has no solutions in the interior of the cylinder Q in the (t, y, z) space. The minimum is attained on the boundary of Q.

Suppose the minimum is attained on the upper boundary $\Gamma_1(t)$. We thus have

$$\bar{y} = \Gamma_1(t). \tag{S. 55}$$

The local slope $\bar{z}$ of the minimal is obtained from the condition

$$S(t, \Gamma_1(t), \bar{z}) = \inf_{-\infty < z < \infty} S(t, \Gamma_1(t), z). \qquad (S.\ 56)$$

If $\bar{z}(t)$ is finite, it satisfies the necessary condition

$$S_z = F_z(t, \Gamma_1(t), z) - W(t, \Gamma_1(t)) = 0. \qquad (S.\ 57)$$

If $\Gamma_1(t) \in U$ and $\bar{z}(t)$ is bounded, we have $\bar{u} \in U_0$. We would like to make one observation concerning this case.

If $\Gamma'(t) - \bar{z}(t) < 0$, the approximating sequence $\{\gamma_n\}$ constructed along the previous lines, i. e., the curve

$$y_i = \bar{y}(t_i + 0); \quad y_i' = \bar{z}(t_i + 0), \quad i = 0, 1, 2, \ldots n,$$

does not belong to the set of admissible lines, since it extends beyond the boundary of B. In this case we should take

$$\tilde{y}_i = \bar{y}(t_i + 0); \quad y_i' = \bar{z}(t_i + 0), \qquad (S.\ 58)$$

where

$$\tilde{y}_i = y_{i-1} + y_{i-1}' \cdot \Delta t_{i-1} = y_n(t_i - 0).$$

The sequence $\{\gamma_n\}$ constructed in this way clearly belongs to B and, in virtue of the continuity of the functional $I(u)$, $u \in U_0$, in the sense of (S. 42), it is also a minimizing sequence.

4. $D(t, y, z) = 0$ everywhere in the cylinder Q. System (S. 52) is equivalent to the single equation

$$f(t, y, z) = 0. \qquad (S.\ 59)$$

This equation contains two independent unknown functions, y and z. Indeed, the class U_0, in general, contains infinitely many $(y,\ z)$ lines satisfying the necessary conditions (S. 52). These lines may include extremals which belong to U. The latter should satisfy, almost everywhere on $[a,\ b]$, the differential equation

$$f(t, y, y') = 0. \qquad (S.\ 60)$$

Such an extremal may have discontinuities of the first kind on any finite set of arbitrarily selected points $t_i \in [a,\ b]$ containing at least one point. On the smooth sections it satisfies the equation

$$y(t) = y^0(t, C_i),$$

where $y^0(t, C_i)$ is a general solution of (S. 60), and C_i is the integral constant for the i-th section, if there exist $C_i \neq C_{i+1}$ such that condition (S. 54) is satisfied at the points t_i, which takes in the present case the form

$$S(t_i, y^0(t_i, C_i), y^{0'}(t_i, C_i)) = S(t_i, y^0(t_i, C_{i+1}), y^{0'}(t_i, C_{i+1})). \qquad (S.\ 61)$$

Thus in general the extremal $\bar{u}$ belonging to the set U is not unique. The only exception is the case when

$$f_z \equiv 0. \qquad\qquad\qquad \text{(S. 62)}$$

The extremal is then unique and is defined by the equation

$$f(t, y) = 0.$$

In virtue of the properties of $F(t, y, z)$ and $W(t, y)$, the function $\bar{y}(t)$ is continuous and smooth everywhere except the end points, provided $f_y(t, y) \neq 0$ everywhere on (a, b).

A more restricted problem can be formulated: find the extremal with the least number of discontinuities on U. It is readily seen that this extremal in our case consists of two smooth branches which are solutions of equation (S. 60) passing through the points (a, a_1) and (b, b_1) and a straight vertical segment $t = t_0$ joining the ends of these branches, where t_0 is some point of the segment $[a, b]$.

This result coincides with Razmadze's result obtained in /13/. Razmadze's condition for the case of a "floating" discontinuity point also can be obtained without difficulty:

$$\left.\begin{array}{l} F(t_0, y_0, y_0') = F(t_0, \bar{y}_0, \bar{y}_0'); \\[4pt] F_{y'}(t_0, y_0, y_0') = F_{y'}(t_0, \bar{y}_0, \bar{y}_0') = 0, \end{array}\right\} \qquad \text{(S. 63)}$$

where $y_0 = y(t_0 + 0)$, $\bar{y}_0 = y(t_0 - 0)$. To this end, it suffices to take $W(t, y) = 0$. in (S. 52) and (S. 54). Then from the second equation in (S. 52) we obtain the second condition in (S. 6), and from (S. 54) we obtain the first equation in (S. 6). Note that Razmadze's conditions both for a fixed and a "floating" discontinuity are valid only when $W = 0$.

The sufficient conditions of extremum, as before, are presented by (S. 49).

5. As we have seen above, if $\bar{y}(t) \in U$ and $\bar{z}(t) = \bar{y}'(t)$ almost everywhere on $[a, b]$, the minimal $\bar{u}_0 \in U$. It is readily seen that in this case $\bar{y}(t)$ almost everywhere satisfies Euler's equation

$$F_y(t, y, y') - \frac{d}{dt} F_{y'}(t, y, y') = 0. \qquad\qquad \text{(S. 64)}$$

To show this, it suffices to differentiate the second equation in (S. 52) and subtract it from the first.

C o n c l u s i o n s. I. Let us summarize the results of this section. If the function $f(t, y, p, 1)$ for $p = 0$, $(t, y) \in B$, exists and is continuous in p, the absolute extremum of the functional (S. 13) is attained on extremals of a special kind (we call them type a extremals), which are fundamentally different from the classical Euler extremals (which we call type b extremals). In distinction from type b extremals, every linear element (y, z) of type a extremals is independent of the other elements and minimizes the function $S(t, y, z)$ for every fixed $t \in [a, b]$. To obtain finite equations of these extremals, we do not have to solve any boundary-value problems: they are obtained directly in the form of the necessary condition (S. 52). An extremal on U may belong either to this set or to a larger set U_0. In the latter case, although $\bar{u} \bar{\in} U$, the solution $\bar{y}(t)$, $\bar{z}(t)$ fully describes a minimizing sequence $\{\gamma_n\} \in U$ (see Definition 3).

II.   The function $f(t, y, p, 1)$ for $p=0$ and any $t, y \in B$ has a discontinuity of the first kind in p.

In this case, no analog of Theorem 3 giving the necessary and sufficient conditions of minimum and providing a complete solution of the problem exists.   We will only derive the necessary conditions of minimum in U_0.

The minimum is attained both on the (y, z) minimals of the type considered before and on the classical Euler–Lagrange extremals.   Let $u \in U_0$.   By Theorem 2 we have

$$I(u) = \sum_{i=0}^{k} \left\{ \int_{\mu_i + 0}^{\mu_{i+1}^{-0}} [F(t, y, z) + W(t, y, \operatorname{sign}(y' - z))] \times \right.$$

$$\left. \times (y' - z)\, dt + \int_{\bar{y}_i}^{y_i} W(\mu_i \xi, \operatorname{sign}(y_i - \bar{y}_i))\, d\xi \right\}. \tag{S. 39}$$

Here μ_i are discontinuity points of $y(t)$.   By (S. 39), the functional $I(u)$ may be treated as depending on a pair of independent functions $y(t)$, $z(t)$. Using expression (S. 39), we can write and investigate the expression for the first variation of $I(u)$ with respect to $y(t)$ and $z(t)$ in the usual classical sense.   The class of admissible functions is further restricted by the requirement of smooth $z(t)$ on the intervals (μ_i, μ_{i+1}).   The discontinuity points μ_i are assumed fixed.

The necessary condition of minimum is

$$\delta I(u) \geqslant 0. \tag{S. 65}$$

The expression of the first variation is

$$\delta I(u) = \sum_{i=0}^{k} \int_{\mu_i + 0}^{\mu_{i+0}^{-0}} [F_y \delta y + F_z \delta z + W_y(y' - z)\delta y +$$

$$+ W\delta(y' - z)]\, dt + \sum_{i=0}^{k} \{W[\mu_i, y_i, \operatorname{sign}(y_i - \bar{y}_i)]\,\delta y_i -$$

$$- W[\mu_i, \bar{y}_i, \operatorname{sign}(y_i - \bar{y}_i) + \delta y_i - \delta \bar{y}_i)]\,\delta y_i\}, \tag{S. 66}$$

where

$$\delta y = y^*(t) - y(t); \quad \delta z = z^*(t) - z(t); \quad \delta y' = y^{*\prime}(t) - y'(t),$$

$y^*(t)$, $z^*(t)$ are the corresponding functions of the line $u^* \in U_0$ which is sufficiently close to first order to the line u.

Since the functional (S. 39) is linear in $y'(t)$, its variation with respect to y' coincides with the increment with respect to $y'(t)$.   Therefore $y^{*\prime}$ is not necessarily close to $y'(t)$ and we may take $\delta y(t) \in U$.   After simple manipulations, the integral term in the expression of the first variation is easily written in the form

$$\delta I(u)_1 = \sum_{i=0}^{k} \int_{\mu_i + 0}^{\mu_{i+1}^{-0}} \{[F_y - W_y(y' - z)]\,\delta y + F_z \delta y' + (W - F_z)\delta(y' - z)\}\, dt.$$

This expression gives $\delta I(u)_1$ as a functional of four independent functions $y(t)$, $z(t)$, $\delta y(t)$ and $\delta(y' - z(t))$.   Since $\delta y(t) \in U$, we regard the functions

$y(t), z(t)$, and $\delta(y'-z(t))$ as given and use (S. 31) and (S. 39) to obtain

$$\delta I(u) = \sum_{i=0}^{k} \int_{\mu_i+0}^{\mu_{i+1}-0} \left\{ [F_y + W_y(y'-z)]\delta y - \int_{c=0}^{\delta y} \frac{d}{dt} F_z(t,\ y(t),\ z(t)\,d\xi \right\} dt +$$

$$+ \sum_{i=0}^{k} \int_{\mu_i+0}^{\mu_{i+1}-0} [W(t,\ y,\ \text{sign}(y'-z+\delta y'-\delta z)) - F_z]\delta(y'-z) \times$$

$$\times dt + \sum_{i=0}^{k} \left[\int_{c=0}^{\delta y_{i+1}} F_z(\mu_{i+1})\,d\xi - \int_{c=0}^{\delta y} F_z(\mu_i)\,d\xi \right] + \sum_{i=0}^{k} [W(\mu_i\ y,\ \text{sign}(y_i-\overline{y}_i)\delta y_i -$$

$$- W(\mu_i,\ \overline{y}_i,\ \text{sign}(y-\overline{y}_i))\delta\overline{y}_i] + \sum_{j=0}^{k_1-k} [(W(t_j,\ y_j,\ \text{sign}(\delta y_j - \delta\overline{y}_j)) - F_z](\delta y_j - \delta\overline{y}_j),$$

where k_1 is the number of discontinuity points of $\delta y(t)$.

Finally we obtain

$$\delta I(u) = \sum_{i=1}^{k} \int_{\mu_i+0}^{\mu_{i+1}-0} \left[F_y - \frac{d}{dt}F_z + W_y(y'-z) \right] \delta y\,dt +$$

$$+ \sum_{i=0}^{k} \int_{\mu_i+0}^{\mu_{i+1}-0} [W(t,\ y,\ \text{sign}(y'-z+\delta y'-\delta z)) -$$

$$- F_z]\delta(y'-z)\,dt + \sum_{i=0}^{k} \{[W(\mu_i,\ y_i,\ \text{sign}(y_i-\overline{y}_i)) - F_z]\delta y_i -$$

$$- [W(\mu_i,\ \overline{y}_i,\ \text{sign}(y_i-\overline{y}_i))] - F_z(\mu_i,\ \overline{y}_i,\ \overline{y})\,\delta\overline{y}_i \} +$$

$$+ \sum_{j=0}^{k_1-k} [W(t_j,\ y_j,\ \text{sign}(\delta y_j - \delta\overline{y}_i)) - F_z(t_j, y_j,\ z_j)](\delta y_j - \delta\overline{y}_j). \qquad \text{(S. 67)}$$

Here δy and $\delta(y'-z)$ are independent variations. (S. 67) leads to the following conclusions.

1. The functional may have a minimum only if the right and the left limits of $f(t,\ y,\ \pm 0,1)$ along the extremal satisfy the condition

$$W(t,\ y,\ \text{sign}\,\varepsilon)\,\varepsilon \geqslant 0. \qquad \text{(S. 68)}$$

This is readily established if we take in (S. 67)

$$\delta y(t) = 0$$

for

$$|t-t_0| \geqslant \eta,\ t_0 \in [a,\ b]$$

and

$$\delta y = \delta y(t_0) \neq 0$$

for

$$|t-t_0| < \eta,\ |\eta| \to 0.$$

For a maximum, this inequality should be reversed. This condition can be regarded as the first necessary condition to be satisfied by the functional if it is to have an extremum on the class U of admissible lines.

2. The extremals of the functional (S. 13) may be of the following types:

a) The functions $y(t)$ and $z(t)$ satisfy the equations

$$\left.\begin{aligned}
S_y &\equiv F_y(t,y,z) - W_y(t,y,\operatorname{sign}(y'-z))z - W_t(t,y) = 0; \\
S_z &\equiv F_z(t,y,z) - W(t,y,\operatorname{sign}(y'-z)) = 0.
\end{aligned}\right\} \tag{S. 69}$$

It is readily seen that these equations lead to zero coefficients before δy and $\delta(y'-z)$. To verify this, it suffices to differentiate the second equation in (S. 69) and subtract it from the first. Equations (S. 69) may be written in a more detailed form

$$\left.\begin{aligned}
S'_y &\equiv F_y - W_y(t,y,1)\,z - W_t(t,y,1) = 0; \\
S_z &\equiv F_z - W(t,y,1) = 0, \quad y'-z > 0;
\end{aligned}\right\} \tag{S. 70}$$

$$\left.\begin{aligned}
S_y &\equiv F_y - W_y(t,y,-1) - W(t,y,-1) = 0; \\
S_z &\equiv F_z - W(t,y,-1) = 0, \quad y'-z < 0.
\end{aligned}\right\} \tag{S. 71}$$

Along these extremals, the first variation is zero, i. e., the extremals are stationary. Moreover, as we see from (S. 69), they satisfy the necessary conditions of minimum of the function $S(t,\,y,\,z)$, but now the function additionally depends on the sign of the difference $y'-z$. These extremals are analogous to the $(y,\,z)$ minimals of the previous section. Any infinitesimally small section of these extremals retains the property of maximum or minimum and is independent of all other elements, but only within the limits consistent with the inequalities in (S. 70) and (S. 71).

All that we have said in the previous section remains valid for these extremals with one reservation: condition (S. 49) is not sufficient in this case for an absolute minimum.

b) The functions $y(t)$ and $z(t)$ are continuous on $(a,\,b)$ and satisfy the conditions

$$\left.\begin{aligned}
&z(t) = y'(t); \quad F_y - \frac{d}{dt}F_z = 0; \\[2mm]
&[W(t,y,\operatorname{sign}\varepsilon) - F_{y'}(t,y,y')] \cdot \varepsilon \geqslant 0, \quad \varepsilon \neq 0; \\[2mm]
&[W(a,y(a),\operatorname{sign}(y(a)+\delta y(a)-a_1)) - F_{y'}(a,y(a),y'(a))]\,\delta y(a) \geqslant 0; \\[2mm]
&[W(b,\bar{y}(b),\operatorname{sign}(b_1-\bar{y}(b)-\delta y(b))) - F_{y'}(b,\bar{y}(b),\bar{y}'(b))]\,\delta\bar{y}(a) \geqslant 0.
\end{aligned}\right\} \tag{S. 72}$$

The first two equations in (S. 72) define $y(t)$ and $z(t)$ and show that a type b extremal on $U(U_0)$ is a continuous differentiable function which satisfies Euler's equation. Every infinitesimal section of these extremals is dependent on the position of the neighboring elements and, in general, does not possess the property of maximum or minimum.

However, unlike the extremals in class C, the extremals in class U should satisfy an additional condition on $(a,\,b)$, namely the first inequality in (S. 72).

The last two inequalities in (S. 72) define extremal boundary conditions for the determination of the integral constants C_1 and C_2. Since $\delta y(a)$ is arbitrary, the first of these inequalities is satisfied if one of the following three conditions holds true:

$$
\begin{aligned}
&1)\ W(a,y(a),1)-F_{y'}(a,y(a),\,y'(a))=0; \\
&\qquad\qquad\qquad y(a)-a_1>0; \\
&2)\ W(a,y(a),1)-F_{y'}(a,y(a),y'(a))=0; \\
&\qquad\qquad\qquad y(a)-a_1<0; \\
&3)\ y(a)=a_1.
\end{aligned}
\qquad\text{(S. 72a)}
$$

The second inequality defines analogous conditions at the right end. Thus, if the solution of Euler's equation is unique for each pair of these boundary conditions, this equation together with (S. 72a) and analogous conditions on the right end may produce nine type b extremals. The first two equalities in (S. 72a) constitute natural boundary conditions; they are meaningful if $y(a)-a_1>0$ or $y(a)-a_1<0$, respectively. The last equality in (S. 72a) follows directly from the first inequality in (S. 72); it shows that the solution of Euler's equation for the boundary conditions $y(a)=a_1$ and $y(b)=b_1$ is also an extremal in U.

c) It follows from (S. 67) that, in addition to type a and type b extremals, there are also mixed extremals consisting of pieces of type a and pieces of type b, with the functions $y(t)$ and $z(t)$ suffering discontinuities at the points μ_i.

In drawing up the expression for the first variation (S. 67), we used the comparison lines u^* whose functions $y(t)$ and $z(t)$ are close to the original $y(t)$ and $z(t)$ everywhere, except at the discontinuity points μ_i. We now enlarge the group of comparison lines by including (y,z) lines with the discontinuity point displaced in an ε-neighborhood of μ_i. Using expression (S. 61) for the functional $I(u)$ and comparing it with the functional along these lines, we obtain, as in the previous section, an additional necessary condition, which should be observed at the discontinuity points μ_i:

$$
S(\mu_i,\,y_i,\,z_i)=S(\mu_i,\,\bar{y}_i,\,\bar{z}_i). \qquad\text{(S. 73)}
$$

Here the pairs of functions $y(t),\,z(t)$ and $\bar{y}(t),\,\bar{z}(t)$ are either one of the solutions of system (S. 69) or the solution of Euler's equation with $z(t)=y'(t)$ and one of the boundary conditions at the discontinuity point μ_i,

$$
\begin{aligned}
&1)\ W(\mu_i,y_i,1)-F_y(\mu_i,y_i,y_i')=0; \\
&\qquad\qquad\qquad y_i-\bar{y}_i>0; \\
&2)\ W(\mu_i,y_i,-1)-F_{y'}(\mu_i,y_i,y_i')=0; \\
&3)\ y_i=\bar{y}_i,
\end{aligned}
\qquad\text{(S. 74)}
$$

considered jointly with one of the analogous conditions at the other end point.

We have thus established that if the left and the right limits of $f(t,\,y,\,\pm0,1)$ exist, and are different from each other, the functional (S. 13) may have extremals of two types, a and b. Type a extremals are analogous to the (y, z) extremals of the previous section, and type b extremals coincide with the extremals in the class of continuous functions on (a, b). There are also mixed extremals consisting of pieces of types a and b. Additional necessary conditions have been derived which the functional must satisfy in order to have an extremum in U, which coincides with a weak local extremum in the class of continuous functions C.

1.4. Special cases

We will now consider the most characteristic particular cases when the condition of existence of $\Phi(\mu_i, y, \bar{y})$ enables us to fix the various discontinuity points of the extremal $\bar{y}(t)$ and to determine its qualitative behavior.

I. The functions $f(t, y, \pm 0,1)$ do not exist anywhere in region B in the (t, y) plane.

The function $\Phi(t, y, \bar{y})$ does not exist for all $t, y, \bar{y} \in B$, $y \neq \bar{y}$. The integral (S.13) exists and is finite only on the continuous curves from U. The minimal, if it exists, belongs to this category. The problem thus reduces to finding a minimum in the class C of continuous functions. Further investigation of this case requires application of the theoretical-functional techniques $(1-4)$ described in the introduction.

An extremal in general depends on two parameters which are determined by the values of the function $y(t)$ at the two ends, $y(a) = a_1$; $y(b) = b_1$. Under certain conditions, this is a smooth curve satisfying the Euler equation.

Functionals with the functions $f(t, y, \pm 0,1)$ existing at a finite number of points $t = \mu_i$ are close to this type. As before, the condition of existence of $\Phi(t, y, \bar{y})$ limits the class of admissible curves to curves from U on which $y(t)$ is continuous everywhere, possibly with the exception of $t = \mu_i$ $(i = 1, 2, \ldots, k)$. The minimal, if it exists, also belongs to this set.

The problem thus reduces to finding an extremum in the class of continuous piecewise-smooth functions and determining the extremal "matching" conditions of the continuous pieces at the verticals $t = \mu_i$. We will confine our analysis of the functionals of this type to the derivation of the "matching" conditions and discussion of some of the corollaries.

These conditions are determined by the properties of the function $f(t, y, p, 1)$ for $p = 0$ on the straight lines $t = \mu_i$.

II. The function $f(t, y, p, 1)$ exists and is continuous on the straight line

$$t = \mu_i, \quad p = 0, \quad \Gamma_2(\mu_i) \leqslant y \leqslant \Gamma_1(\mu_i).$$

Consider one such value $t = \mu \in (a, b)$. The function W is independent of sign Δy, i. e.,

$$W(\mu, y, 1) = W(\mu, y, -1)$$

and we may write

$$\Phi(\mu, y, \bar{y}) = \Phi(\mu, y, c) - \Phi(\mu, \bar{y}, c), \tag{S. 75}$$

where c is a constant.

By (S. 18) and (S. 75),

$$
\left.\begin{aligned}
I(u) &= I^{(1)} + I^{(2)}; \\[4pt]
I^{(1)} &= \int_a^{\mu} F(t,y,y')\,dt - \Phi(\mu,\overline{y}(\mu),c); \\[4pt]
I^{(2)} &= \int_{\mu}^b F(t,y,y')\,dt + \Phi(\mu,y,c).
\end{aligned}\right\}
\tag{S. 76}
$$

$I(u)$ thus separates into two independent functionals. Therefore, its extremal is made up of two continuous pieces, which are extremals of the independent functionals (S. 76), joined by a vertical segment $t=\mu_0$. Since $\Phi(\mu, y, c)$ is continuous together with its derivative, these pieces at $t=\mu_0$ satisfy natural boundary conditions of the form

$$
F_{y'} - \frac{d}{d\overline{y}_0}\,\Phi(\mu,y,c)=0; \qquad F_{y'} - \frac{d}{dy_0}\,\Phi(\mu,\overline{y},c)=0,
$$

or

$$
\left.\begin{aligned}
F_{y'}(\mu_0,y_0,y_0') - W(\mu_0,y_0) &= 0; \\[4pt]
F_{y'}(\mu_0,\overline{y}_0,\overline{y}_0') - W(\mu_0,\overline{y}_0) &= 0.
\end{aligned}\right\}
\tag{S. 77}
$$

It is remarkable that each piece of the extremal depends only on the position of one of its end points, and is independent of the position of the other end. If the continuous pieces of the extremal are piecewise-smooth, they satisfy Euler's equation with boundary conditions (S. 13) and (S. 77) and Erdmann—Weierstrass conditions at the corner points. In the particular case $W(t_0, y)=0$, conditions (S. 77) coincide with Razmadze's conditions for a fixed discontinuity point /13/.

The results are readily generalized to the case of n discontinuity points. In this case, the extremal consists of $n+1$ continuous pieces, and each i-th piece of the extremal coincides with the extremal of the functional

$$
I_i = \int_{\mu_{i-1}}^{\mu_i} F(t,y,y')\,dt + \int_c^{y_{i-1}} W(\mu_{i-1},\xi)\,d\xi - \int_c^{y_i} W(\mu_i,\xi)\,d\xi.
\tag{S. 78}
$$

Each of these extremals is a curve dependent on two parameters. These parameters are determined from the natural boundary conditions at the discontinuity points μ_{i-1} and μ_i:

$$
\left.\begin{aligned}
F_{y'}(\mu_{i-1}, y_{i-1}, y_{i-1}') - W(t_{i-1},y_{i-1}) &= 0; \\[4pt]
F_{y'}(\mu_i,\overline{y}_i,\overline{y}_i') - W(\mu_i,\overline{y}_i) &= 0, \quad i=1,2,\ldots,n-1.
\end{aligned}\right\}
\tag{S. 79}
$$

For the first and the last piece of the extremal, the two parameters are determined from the conditions

$$
y(a)=a_1; \quad F_{y}'(\mu_0,\overline{y}_0,\overline{y}_0') - W(\mu_0,\overline{y}_0)=0
\tag{S. 80}
$$

and

$$
\overline{y}(b)=b_1; \quad F_{y'}(\mu_n,y_n,y_n') - W(\mu_n,y_n)=0.
\tag{S. 81}
$$

If the continuous sections of the extremal function are piecewise-smooth, they satisfy Euler's differential equation for the functional (S.13) and the Erdmann—Weierstrass conditions at the corner points. A remarkable feature of the extremals is that, with the exception of the pieces adjoining the end points, they are independent of the values of the function $y(t)$ at the ends of $[a, b]$.

III. The functions $f(\mu_i, y, p, 1)$ are discontinuous in p for $p=0$, but the limits $f(\mu_i, y, \pm 0, 1)$ exist and are continuous for $y \in [\Gamma_2(\mu_i), \Gamma_1(\mu_i)]$.

Consider one such point

$$t = \mu_0 \in (a, b).$$

By (S. 18)

$$I(u) = \int_0^{\mu_0} F(t, \overline{y}(t), \overline{y}'(t))\, dt + \Phi(\mu_0, y_0, \overline{y}_0) + \int_{u_0}^{b} F(t, y, y')\, dt. \qquad \text{(S. 82)}$$

The function $\Phi(\mu_0, y_0, \overline{y}_0)$ in this case no longer can be written in the form (S. 75) and its derivatives Φ_{y_0} and $\Phi_{\overline{y}_0}$ are discontinuous at $y_0 - \overline{y}_0 = 0$. These derivatives are expressed by the following relations:

$$\left.\begin{aligned}
\Phi_{y_0} &= \begin{cases} W(\mu_0, y_0, 1) & \text{for } y_0 - \overline{y}_0 > 0; \\ W(\mu_0, y_0, -1) & \text{for } y_0 - \overline{y}_0 < 0; \end{cases} \\
\Phi_{\overline{y}_0} &= \begin{cases} -W(\mu_0, \overline{y}_0, 1) & \text{for } y_0 - \overline{y}_0 > 0; \\ -W(\mu_0, \overline{y}_0, -1) & \text{for } y_0 - \overline{y}_0 < 0. \end{cases}
\end{aligned}\right\} \qquad \text{(S. 83)}$$

We will only consider the case when the continuous pieces of the extremal are smooth. Then the first variation of the functional (S. 83) may be written in the form

$$\delta I(u) = \int_a^{\mu_0} \left[F_y - \frac{d}{dt} F_{y'} \right] \delta \overline{y}\, dt + F_{y'}(\mu_0, \overline{y}_0, \overline{y}_0')\, \delta \overline{y}_0 +$$

$$+ \int_{\mu_0}^{b} \left[F_y - \frac{d}{dt} F_{y'} \right] \delta y\, dt - F_{y'}(\mu_0, y_0, y_0')\, \delta y_0 + \delta \Phi(\mu_0, y_0, \overline{y}_0), \qquad \text{(S. 84)}$$

where

$$\delta \Phi(\mu_0, y_0, \overline{y}_0) = \begin{cases}
W(\mu_0, y_0, 1)\, \delta y_0 - W(\mu_0, \overline{y}_0, 1)\, \delta \overline{y}_0 \\
\qquad\qquad \text{при } y_0 - \overline{y}_0 > 0 \\
W(\mu_0, y_0, -1)\, \delta y_0 - W(\mu_0, \overline{y}_0, -1)\, \delta \overline{y}_0 \\
\qquad\qquad \text{for } y_0 - \overline{y}_0 < 0
\end{cases} \qquad \text{(S. 85)}$$

$$\left.\begin{aligned}
W(\mu_0, y_0, 1)(\delta y_0 - \delta \overline{y}_0), &\quad \text{if } \delta y_0 - \delta \overline{y}_0 > 0 \\
W(\mu_0, y_0, -1)(\delta y_0 - \delta \overline{y}_0), &\quad \text{if } \delta y_0 - \delta \overline{y}_0 < 0
\end{aligned}\right\} \text{for } y_0 - \overline{y}_0 = 0.$$

The necessary condition of a minimum of the functional $I(u)$ is

$$\delta I(u) \geqslant 0. \qquad \text{(S. 86)}$$

This condition is satisfied if both pieces of the extremal satisfy Euler's equation

$$F_y - \frac{d}{dt} F_{y'} = 0, \qquad \text{(S. 87)}$$

and one of the following conditions is satisfied at the discontinuity point:

$$\left.\begin{aligned}
F_{y'}(\mu_0, y_0, y_0') - W(\mu_0, y_0, 1) = 0; \\
F_{y'}(\mu_0, \bar{y}_0, \bar{y}_0') - W(\mu_0, \bar{y}_0, +1) = 0;
\end{aligned}\right\} \qquad \text{(S. 88)}$$

$$y_0 - \bar{y}_0 > 0;$$

$$\left.\begin{aligned}
F_{y'}(\mu_0, y_0, y_0') - W(\mu_0, y_0, -1) = 0; \\
F_{y'}(\mu_0, \bar{y}_0, \bar{y}_0') - W(\mu_0, \bar{y}_0, -1) = 0;
\end{aligned}\right\} \qquad \text{(S. 89)}$$

$$y_0 - \bar{y}_0 < 0;$$

$$[F_{y'}(\mu_0, y_0, y_0') - W(\mu_0, y_0, 1)][F_{y'}(\mu_0, y_0, y_0') - W(\mu_0, y_0, -1)] \leqslant 0; \qquad \text{(S. 90)}$$

$$y_0 - \bar{y}_0 = 0.$$

Thus, if Euler's equation (S. 87) is uniquely solvable for the boundary conditions (S. 14), (S. 88), (S. 89), and (S. 90), the functional (S. 13) may have three extremals in this case which satisfy one of the three conditions above at the point where the limit $f(t, y, \pm 0, 1)$ exists. Two of these extremals, specifically those satisfying conditions (S. 88) and (S. 89), are discontinuous at this point. Unlike case II, the continuous pieces of these extremals are no longer independent of one another, since their end points on the line $t = \mu_0$ are interlinked by the inequalities in (S. 88) and (S. 89); however, between the limits compatible with these inequalities, the different pieces are still independent. Condition (S. 90) shows that besides discontinuous extremals, a continuous extremal may also exist in this case, satisfying Euler's equation and the boundary conditions (S. 14).

If there are n points μ_i $(i = 0, 1, 2, \ldots, n-1)$ on the segment $[a, b]$ at which the limits $W(\mu_i, y, 1)$ and $W(\mu_i, y, -1)$ exist but are different, the necessary extremum conditions are satisfied by any function $y(t)$ consisting of continuous pieces which satisfy Euler's equation, conditions (S. 14) at the ends, and one of the conditions (S. 88), (S. 89), (S. 90) at the points μ_i $(i = 0, 1, \ldots, n-1)$. In particular, one of such functions is the continuous function $y(t)$ satisfying Euler's equation on $[a, b]$ with boundary conditions (S. 14).

We considered the most characteristic cases when the existence of the integrals $\Phi(\mu_i, y_i, \bar{y}_i)$ provides some indication of the qualitative behavior of the minimal and fixes the position of the discontinuity points. In addition to cases I—III, there may be mixed cases when for some $t = \mu_1$ the conditions of case II apply, whereas for $t = \mu_2$ the conditions of case III. In some cases, only one of the limits $f(\mu, y, \pm 0, 1)$ may exist on the vertical $t = \mu_0$, e. g., $f(\mu, y, +0, 1)$. In this case, the condition of existence of $\Phi(\mu, y_0, \bar{y}_0)$ shows that the minimal $\bar{y}(t)$ at $t = \mu_0$ may only display a positive jump satisfying (S. 88) or not jump at all, i. e., $y_0 - \bar{y}_0 = 0$.

1.5. Examples

E x a m p l e 1. Consider the extremum of the linear functional

$$I = \int_a^b [P(t,y)+Q(t,y)\,y']\,dt, \qquad y(a)=a_1,\; y(b)=b_1. \tag{S.91}$$

This example is of independent interest because of numerous applications.
We have

$$f(t,y,0,1)=\lim_{p\to 0} p\left[P+Q\frac{1}{p}\right]=Q(t,y);$$
$$W(t,y)=Q(t,y);$$
$$S=P(t,y)-\int_{c=\text{const}}^{y} Q_t(t,\xi)\,d\xi. \tag{S.92}$$

For the functional (S.91) to attain an extremum on the function $y=y(t)$,
it is necessary and sufficient for the function of a single variable $S(t,\,y)$ to
have an appropriate extremum for every fixed $t \in [a,\,b]$. The equation of
the extremal is

$$S_y \equiv P_y(t,\,y)-Q_t(t,\,y)=0,\; t \in (a,\,b). \tag{S.93}$$

We thus obtained the necessary and sufficient condition of extremum of
a linear functional (S.91) in the class of curves with vertical segments.
The extremal, in general, consists of vertical segments $t=a$ and $t=b$ and
the curve (S.93) and belongs to the set U.

Equation (S.93) also can be obtained by the method of variations, in the
form of degenerate Euler's equation. Using Green's theorem, we can
derive the sufficient conditions for a strong local minimum on its solution
$y^0(t)$. It coincides with the condition of local minimum of $S(t,\,y)$ in the
neighborhood of y^0 for every fixed t. If equation (S.93) has several solutions,
the absolute minimal consists of pieces of these solutions and pieces of
the boundary, joined by vertical segments.

The construction of the absolute minimal from these pieces by the
method of variations combined with Green's theorem is not a simple
process. Our method provides an attractively simple solutions to this
problem: using (S.92) we construct the function $S(t,\,y)$ and for every fixed
$t \in [a,\,b]$ find the value of $\bar{y}(t)$ on which $S(t,\,y)$ attains it least value on
the segment $\Gamma_2(t) \leqslant y \leqslant \Gamma_1(t)$.

The fact that S is independent of z signifies that there are infinitely
many minimals $\bar{u}$ in U_0. This category includes any $(y,\,z)$ line $\bar{u} \in U_0$ with
a zero closeness function $\bar{y}(t)$ constructed by the above method for an
arbitrary (finite) local slope $z(t)$. In other words, any sequence of polygonal
lines $\{\gamma_n\} \subset U$ satisfying the condition

$$y_i=\bar{y}(t_i+0),\; |y'_i|<M,\; i=0,\,1,\dots,\,n-1,$$

where $M>0$ is any number, is a minimizing sequence.

Let $P=Q=ty$. The equation of the extremal is $t-y=0$, $t\in(a,\ b)$. Further, $S=ty-\frac{y^2}{2}$. Along the extremal $S=\frac{t^2}{2}$ we have

$$\Delta S=ty-\frac{y^2}{2}-\frac{t^2}{2}=-\frac{1}{2}(t-y)^2<0.$$

Therefore, along a polygonal line consisting of the vertical segments $t=a$ and $t=b$ and the straight line (S. 93) the functional attains its absolute minimum. The extremal is unique.

Example 2. (Razmadze). Consider the extremum of the functional

$$I=\int_a^b \sin(yy')\,dt;\quad y(a)=a_1;\ y(b)=b_1. \tag{S. 94}$$

The following conditions are satisfied:

$$f(t,y,0,1)=\lim_{p\to 0} p\sin\left(y\ \frac{1}{p}\right)\ \text{exists and is continuous;}$$

$$W(y)=0;$$

$$S(t,y,z)=F(t,y,z)=\sin(yz);$$

$$S_{max}=1;\ S_{min}(y,z)=-1$$

for any y and z, so that we obtain the following equations for the two families of extremals in U_0:

$$yz=\frac{\pi}{2}+2k\pi,\quad yy'=-\frac{\pi}{2}+2k\pi, \tag{S. 95}$$

$$k=0,\pm1,\pm2,\dots$$

The equations of the pieces of the extremals belonging to U are

$$yy'=\frac{\pi}{2}+2k\pi,\quad yy'=-\frac{\pi}{2}+2k\pi.$$

Integration gives

$$y^2=(2k+1)\pi t+C_i;\quad y^2=(2k-1)\pi t+C_i. \tag{S. 96}$$

$$k=0,\pm1,\pm2,\dots$$

Since the function S is independent of C_i, any extremal may have discontinuities at any point $t_i\in[a,\ b]$.

The functional (S. 95) thus has two families of extremals in U, consisting of pieces which satisfy the first and the second equation in (S. 96), with arbitrary C_i, and have any number of discontinuity points with arbitrary abscissas. On the first of these two families the functional (S. 94) attains its absolute maximum, and on the second family its absolute minimum.

Example 3. Minimize the functional

$$\left.\begin{array}{c} I(y)=\int_a^b |y|\sqrt{1+y'^2}\,dt; \\[2mm] y(a)=a_1>0,\ y(b)=b_1>0 \end{array}\right\} \tag{S. 97}$$

(the problem of the least surface of revolution). Here

$$W(t, y, \pm 1) = \lim_{p \to \pm 0} |y| \sqrt{1 + \frac{1}{p^2}} \, \frac{p}{m} = \pm |y|,$$

where W depends on the sign of p. Therefore, the functional (S. 97) corresponds to type II of the general case.

Further,

$$W(y, \text{ sign } \varepsilon) \cdot \varepsilon = |y \cdot \varepsilon| \geqslant 0,$$

i. e., the functional (S. 97) satisfies the necessary condition of the existence of extremum (S. 68).

Let us find a type a extremal. We have

$$\left. \begin{array}{l} S(t,y,z) \equiv |y| \sqrt{1+z^2} - |y| z \, \text{sign}(y'-z); \\[2mm] S(y,z) \equiv |y| \sqrt{1+z^2} - z \, \text{sign}(y'-z). \end{array} \right\} \tag{S. 98}$$

We use the necessary and sufficient condition of a local extremum (S. 49), which is also valid for functionals with a discontinuous derivative F_y, such as (S. 97). We have

$$S_{\text{min}}(y, \, z) = S(0, \, z) = 0.$$

The equation of the extremals is

$$y(t) \equiv 0; \, a < t < b \tag{S. 93}$$

for any z. Since z is arbitrary, there exist infinitely many type a extremals in U_0. This is a reflection of the fact that for any curve $y^*(t) \neq 0$, $|y^*(t)| < \eta$, where η is sufficiently small, there is always a polygonal line with arbitrary section slopes

$$y_i = z(t_i) \quad \text{and} \quad y_i = 0 \ (i=0, \, 1, \, \ldots, \, n),$$

where n is finite, such that $I(\gamma_n) < I(y^*)$. Among these extremals, there is one which belongs to U, specifically

$$y(t) \equiv 0; \, z(t) = y'(t) \equiv 0; \, a < t < b. \tag{S. 99}$$

This extremal is unique. It is continuous on $(a, \, b)$ and has discontinuities at $t=a$ and $t=b$. The surface of revolution of this curve comprises two disks of radii a_1 and b_1, joined by a tube of zero radius. The minimum value of the functional (S. 97) is obtained from (S. 40):

$$I_{\text{min}} = \int_a^b S_{\text{min}}(y,z)\,dt + \int_{c=0}^{b_1} |y| \, \text{sign}(b_1 - \bar{y}(b))\,dy -$$

$$- \int_{c=0}^{a_1} |y| \, \text{sign}(y(a) - a_1)\,dy = \frac{1}{2}(a_1^2 + b_1^2).$$

Let us now find a type b extremal. For a curve to be such an extremal in
U, it is necessary and sufficient for it to be an extremal in the class $\bar{C}$ of
continuous functions on (a, b), and for the functional to satisfy the first inequality in (S. 72) along this curve.
Let us check this condition.

$$[W(t, y, \text{sign } \varepsilon) - F_{y'}]\varepsilon \equiv |y|\left[|\varepsilon| - \varepsilon \frac{y'}{\sqrt{1+y'^2}}\right] \geqslant 0.$$

This condition is satisfied identically, and therefore all extremals in $\bar{C}$
are at the same time type b extremals in U. The extremals in $\bar{C}$ satisfy
Euler's equation, whose solution in this case is described by the catenary

$$y = C_1 \text{ch} \frac{t - C_2}{C_1}, \tag{S. 100}$$

where C_1 and C_2 are integral constants. These constants are determined
from boundary conditions (S. 72), which in this case take the form

$$\left. \begin{array}{l} y(a) = 0; \quad \bar{y}(b) = 0; \\[2mm] y(a) = a_1; \quad \bar{y}(b) = b_1. \end{array} \right\} \tag{S. 101}$$

Three pairs of boundary conditions (S. 101), together with (S. 100), yield
the extremals b, b_1, b_2, which are shown in Figure S. 2. The fourth pair
$y(a) = \bar{y}(b) = 0$ when inserted in (S. 100) leads to equations which are unsolvable
for C_1 and C_2. This is a reflection of the existence of an extremal which is
not a catenary, i. e., does not satisfy Euler's equation. This is the previously
considered polygonal line $ACDB$, a type a extremal, which is continuous and
differentiable on (a, b).

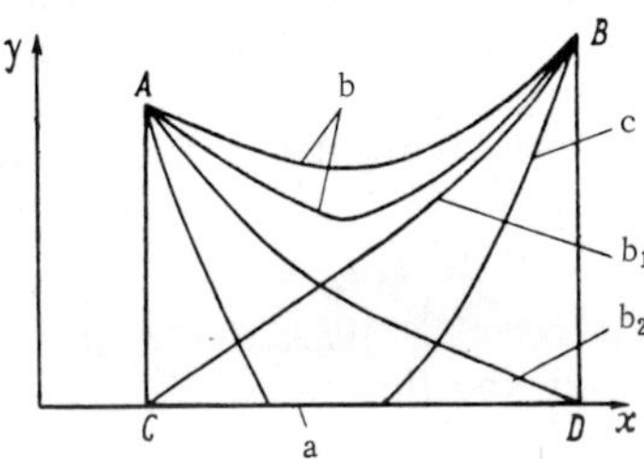

FIGURE S. 2

Finally, let us find mixed type extremals. Since the function $S(y, z,$
$\text{sign}(y'-z))$ is everywhere zero on the t axis and is independent of z and,
moreover, $y \neq 0$ for all $S > 0$, condition (S. 73) is satisfied if and only if

$$y_i = \bar{y}_i = 0.$$

Thus, in addition to type a and b extremals, there are infinitely many
mixed type extremals (type c extremals) in Figure S. 2. However, only two

extremals satisfy the sufficient conditions of minimum: the type a extremal
by (S. 49) and the type b extremal which is continuous on $[a, b]$. The absolute
minimum is attained on one of these curves, depending on the relative
position of the points A and B.

Example 4. Find the extremal in U of the functional

$$I(u)=\int_{-1}^{+1} t^2 y'^2 dt;$$

$$y(-1)=-1; \quad y(+1)=\pm 1. \qquad (S. 102)$$

We have $0 \leqslant I(u) < \infty$, so that we need only consider a minimum. The
function $W(t, y, \pm 1)$ does not exist anywhere on $[-1, 1]$, except at the point
$t=0$, where

$$W(0, y, 1) \equiv W(0, y, -1) \equiv 0,$$

i. e., we are dealing with a type II functional. Therefore, the extremal of
functional (S. 102) consists of two continuous pieces joined at $t=0$ by a
vertical segment. The left piece is the extremal of the functional

$$I_1=\int_{-1}^{0} t^2 y'^2 dt + \int_{c}^{\bar{y}_0} W(0,y) dy = \int_{-1}^{0} t^2 y'^2 dt$$

for $y(-1)=-1$ and the right piece is the extremal of the functional

$$I_2=\int_{0}^{1} t^2 y'^2 dt + \int_{c}^{y_0} W(0,y) dy = \int_{0}^{1} t^2 y'^2 dt$$

for $y(1)=1$.

Each of these functionals if positive definite and vanishes for $y'=0$.
Hence, their minimum is attained on straight lines parallel to the t axis
which pass through the points $(-1, -1)$ and $(1, 1)$, respectively. The
extremal u_0 consists of these straight lines and a joining straight segment
$t=0$,

$$I(u_0)=0.$$

Example 5. Minimize the functional

$$I=\int_{a}^{b} [y^2(1+y')+\sin y'] dt;$$

$$y(a)=a_1; \quad y(b)=b_1. \qquad (S. 103)$$

We have

$$f(t,y,\pm 0,1)=\lim_{p \to \pm 0} \frac{1}{p} [y^2(1+p)+\sin p]=y^2,$$

i. e., $f(t, y, 0, 1)^2$ exists and is continuous. Moreover,

$$W = y^2;$$

$$S = y^2 + \sin z;$$

$$S_{\min}(y,z) = S\left(0, 2\pi m - \frac{\pi}{2}\right) = -1.$$

$$m = 0, 1, 2, \ldots$$

The functional (S. 103) has countably many minimals $\{\bar{u}_m\} \subset U_0$. They all have a common zero closeness line

$$\bar{y}(t) \equiv 0; \; a < t < b \qquad \qquad \text{(S. 104)}$$

with different local slopes

$$\bar{z}_m(t) = 2\pi m - \frac{\pi}{2}. \qquad \qquad \text{(S. 105)}$$

This set contains no lines from U.

§ S. 2. FUNCTIONAL ON THE SET OF FUNCTIONS WITH A BOUNDED DERIVATIVE

The above results can be applied to the fundamental problem of minimizing the functional

$$I(u) = \int_a^b F(t, y, p)\, dt \qquad \qquad \text{(S. 106)}$$

under the constraint

$$y' = g(t, y, p); \qquad \qquad \text{(S. 107)}$$

$$|p| \leqslant 1; \qquad \qquad \text{(S. 108)}$$

$$y(a) = a_1; \; y(b) = b_1. \qquad \qquad \text{(S. 109)}$$

It is assumed that $F(t, y, p)$ and $g(t, y, p)$ are continuous and have continuous partial derivatives for any t and y for $|p| \leqslant 1$. It is moreover assumed that for these t, y, p

$$\frac{\partial g}{\partial p} > 0 \qquad \qquad \text{(S. 110)}$$

and the functions $g(t, y, 1)$ and $g(t, y, -1)$ retain a constant sign.

Our problem thus can be stated as follows: among the pairs of functions $y(t)$, $p(t)$ satisfying conditions (S. 107), (S. 108), (S. 109), find a pair $\bar{y}(t)$, $\bar{p}(t)$ on which the functional (S. 106) attains its minimum value. If $y(t)$ is given, equation (S. 107) uniquely defines the function $p(t) = g^{-1}(t, y(t), y'(t))$. The problem is therefore equivalent to minimizing the functional

$$I(y(t)) = \int_a^b \widetilde{F}(t,y,y')\,dt \tag{S.111}$$

$$y(a) = a_1; \quad y(b) = b_1$$

on the set of functions $y(t)$ with a bounded derivative

$$g(t, y, -1) \leqslant y'(t) \leqslant g(t, y, 1). \tag{S.112}$$

Here

$$\widetilde{F}(t, y, y') = F[t, y, g^{-1}(t, y, y')]. \tag{S.113}$$

Let us elucidate some properties of the set of comparison lines, which is designated U^p:

1) By (S.107), (S.108), and (S.110), the function $y(t)$ is continuous along the line $u \in U^p$ and $p(t)$ (and hence $y'(t)$) is bounded and may have discontinuities of the first kind.

2) The line $u \in U^p$ belongs to a closed simply connected region G in the (t, y) plane whose upper boundary $y = \Gamma_2(t)$ corresponds to the solutions of the equations

$$y' = g(t, y, 1) \quad \text{and} \quad y' = g(t, y, -1),$$

respectively, passing through the points $A(a, a_1)$ and $B(b, b_1)$ and the lower boundary $y = \Gamma_1(t)$ to the solution of the equation $y' = g(t, y, 1)$ passing through B and the solution of the equation $y' = g(t, y, -1)$ passing through A.

3) We impose an additional restriction on U^p, namely that $p(t)$ is piecewise-smooth and may only have a finite number of discontinuities of the first kind, and consequently may only contain a finite number of sections with $p = 1$ or $p = -1$.

Let

$$\left. \begin{array}{l} y = \varphi(t, \tau); \\ y = \psi(t, T) \end{array} \right\} \tag{S.114}$$

be the general integrals of the equations

$$y' = g(t, y, 1)$$

and

$$y' = g(t, y, -1),$$

respectively. Here τ and T are integral constants. We have

$$\left. \begin{array}{l} \varphi_t = g(t, y, 1); \\ \psi_t = g(t, y, -1). \end{array} \right\} \tag{S.115}$$

In virtue of the above properties of the function $g(t, y, p)$, only curve $y = \varphi(t, \tau)$ or $y = \psi(t, T)$ passes through each point t_0, y_0 in G, i.e., to every point (t_0, y_0) corresponds a single pair of values τ and T.

Consider a pair of piecewise-smooth functions $y(t) \in U^p$ and $q(t)$, where $|q| \leqslant 1$. We construct the following sequence of polygonal lines $\{\gamma_n\} \in U^p$: the segment $[a, b]$ is partitioned into n intervals by the points

$$a = t_0 < t_1 < \ldots < t_{n-1} < t_n = b.$$

We take

$$y_n(t_i) = y(t_i) \tag{S.115a}$$

and through each point $(t_i, y(t_i))$ pass a pair of curves $y = \varphi(t, \tau_i)$ and $y = \psi(t, T_i)$ which together with the pair $y = \varphi(t, \tau_{i+1})$, $y = \psi(t, T_{i+1})$ on (t_i, t_{i+1}) define a region G_i of the admissible values of $y_n(t)$, which is constructed along the same lines as the region G. For all (t_i, t_{i+1}) we take

$$p_n = q(t) \tag{S.116}$$

and $y_n(t)$ is correspondingly set equal to the solution of the equation $y' = g(t, y, q(t))$ for those $t \in (t_i, t_{i+1})$ where this solution belongs to G_i, and

$$y_n(t) = \varphi(t, \tau_{i+1}) \quad \text{or} \quad y_n(t) = \psi(t, T_{i+1}) \tag{S.117}$$

for all other $t \in (t_i, t_{i+1})$. The choice of equality (S.117) is determined by the condition of continuity of $y_n(t)$.

We say that the sequence $\{\gamma_n\}$ approximates to the line $u \in U^p$ or $\{\gamma_n\} \underset{n \to \infty}{\to} u$ and correspondingly $\{y_n(t)\} \underset{n \to \infty}{\to} u$ if

$$y'(t) = g(t, y, q) \tag{S.118}$$

almost everywhere on $[a, b]$, and for $n \to \infty$,

$$|t_{i+1} - t_i|_{\max} \to 0. \tag{S.119}$$

If the pair y, q does not satisfy (S.118), we say that $\{\gamma_n\}$ approximates to a (y, q) line $u_0 \in U_0^p$, or $\{\gamma_n\} \to u_0$; $y(t)$ is the zero closeness function of the line u_0, and $q(t)$ is a local value of the parameter p.

We take

$$I(u_0) = \lim_{\{\gamma_n\} \to u_0} I(u_n). \tag{S.120}$$

Theorem 1. The functional $I(u_0)$, $u_0 \in U_0^p$, exists and may be represented in the form

$$I(u_0) = \int_a^b \{F(t, y, \pm 1) + [F(t, y, q) - F(t, y, \pm 1)]\} \times$$
$$\times \frac{y' - g(t, y, \pm 1)}{g(t, y, q) - g(t, y, \pm 1)} dt, \tag{S.121}$$

where

$$\pm 1 = \mathrm{sign}[y' - g(t,\, y,\, q)].$$

We have

$$I(\gamma_n) = \sum_{i=0}^{n-1} \{ F(t_i, y_i, q_i)(\bar{t}_i - t_i) + F[t_i, y_i, \mathrm{sign}(y_{i+1} -$$
$$- y(\bar{t}_i))](t_{i+1} - \bar{t}_i)\} + o(\Delta t_i), \tag{S.122}$$

where

$$y_i = y(t_i); \quad q_i = q(t_i); \quad \Delta t_i = t_{i+1} - t_i,$$

$\bar{t}_i \in (t_i,\, t_{i+1})$ is the abscissa of the point where the straight line

$$y = y_i + g(t_i,\, y_i,\, q_i)\,(t - t_i) \tag{S.123}$$

meets the boundary of G_i, i. e., intersects the curve $y = \varphi(t,\, \tau_{i+1})$ if $y_{i+1} - y(\bar{t}_i) > 0$ and the curve $y = \psi(t,\, T_{i+1})$ if $y_{i+1} - y(\bar{t}_i) < 0$.

We have

$$\left.\begin{aligned}
\frac{\bar{t}_i - t_i}{\Delta t_i} &= \frac{y'(t_i) - g(t_i, y_i, \pm 1)}{g(t_i, y_i, q_i) - g(t_i, y_i, \pm 1)} + o(\Delta t_i); \\[4pt]
\mathrm{sign}[y_{i+1} - y(\bar{t}_i)] &= \mathrm{sign}[y_{i+1} - y_i - g(t_i, y_i, q_i) \times \\[2pt]
&\times \Delta t_i] = \mathrm{sign}[y'(t_i) - g(t_i, y_i, q_i) + o(\Delta t_i)].
\end{aligned}\right\} \tag{S.124}$$

For sufficiently small Δt_i,

$$\mathrm{sign}[y_{i+1} - y(\bar{t}_i)] = \mathrm{sign}[y'(t_i) - g(t_i,\, y_i,\, q_i)], \tag{S.125}$$

if

$$y'(t_i) \neq g(t_i,\, y_i,\, q_i).$$

From (S.122), using (S.124) and (S.125), we find

$$I(\gamma_n) = \sum_{i=0}^{n-1} \{[F(t_i, y_i, q_i) - F(t_i, y_i, \pm 1)]\} \times$$

$$\times \frac{y_i' - g(t_i, y_i, \pm 1)}{g(t_i, y_i, q_i) - g(t_i, y_i, \pm 1)} + F(t_i, y_i, \pm 1)\Big\} \Delta t_i + o(\Delta t_i). \tag{S.126}$$

The argument ± 1 in these expressions stands for

$$\mathrm{sign}[y'(t_i) - g(t_i,\, y_i,\, q_i)].$$

In virtue of the properties of $F(t,\, y,\, q)$, $y(t)$, and $q(t)$, the function

$$[F(t, y, q) - F(t, y, \pm 1)]\, \frac{y'(t) - g(t, y(t), \pm 1)}{g(t, y(t), q(t)) - g(t, y(t), \pm 1)} + F(t,\, y(t),\, \pm 1) \tag{S.127}$$

is bounded and continuous almost everywhere on $[a, b]$. Therefore, by the Lebesgue theorem, it is Riemann-integrable. The limit (S. 120), where $I(\gamma_n)$ is defined by (S. 126), therefore exists, is equal to (S. 121), and does not depend on the choice of the sequence $\{\gamma_n\} \to u_0$. Q. E. D.

Corollary 1. If $y' = g(t, y, q)$, the integrand in (S. 121) is equal to $F(t, y, q)$ and, therefore, definitions (S. 106) and (S. 120) coincide on U^p.

Corollary 2. Since y' enters the integrand in (S. 121) in linear form for $y' - g(t, y, q) \neq 0$, and does not appear in the integrand at all for $y' - g(t, y, q) = 0$, the functional (S. 121) is continuous on U_0^p in the sense that for any given $\varepsilon > 0$ there exists $\delta > 0$ such that

$$|I(u) - I(\bar{u})| < \varepsilon, \tag{S. 128}$$

if $|y - \bar{y}| < \eta$, $|q - \bar{q}| < \delta$ almost everywhere on $[a, b]$. According to the lemma, the minimals in U^p and U_0^p coincide, and therefore instead of minimizing the functional in U^p we can minimize it in the class U_0^p of (y, q) lines, using expression (S. 121) for $I(u_0)$. However, expression (S. 121) in its original form is not particularly convenient for the determination of the absolute minimum, since it contains the derivative of the zero closeness function y'. We will adopt a different approach and try to establish a relationship between this problem and the problem of minimum of $I(u_0)$, $u_0 \in U_0$. To this end, we use (S. 31) and Theorem 3. Along any line $u \in U^p$, the equation $y(t) = \varphi(t, \tau)$ defines a piecewise-smooth function $t = t(\tau)$ which has discontinuities of the first kind at the points $\tau = \mu_i$ $(i = 1, 2, \ldots, n)$ corresponding to the sections $p(t) \equiv 1$ of the line u, if any. Since each τ corresponds to a single t, the variable τ can be chosen so that $t(\tau)$ is an increasing function, i. e.,

$$\frac{dt}{d\tau} > 0 \tag{S. 129}$$

on the smooth sections and

$$t(\tau + 0) - t(\tau - 0) > 0 \tag{S. 130}$$

at the discontinuity points.

The functional $I(u)$ may be considered on the set of the functions $y(t)$, and not on the set of $y(\tau)$, if we make the following substitution of variables in (S. 106):

$$t = \varphi_1(\tau, y);$$

$$dt = \frac{\varphi_{1\tau} \cdot g(\varphi_1, y, 1)}{g(\varphi_1, y, 1) - g(\varphi_1, y, p)} \, d\tau, \tag{S. 131}$$

where $\varphi_1(\tau, y)$ is the inverse of $\varphi(\tau, t)$. Since $g(t, y, 1)$ maintains a constant sign, it exists and is continuous and differentiable.

We have

$$I(u) = \sum_{i=1}^{n-1} \int_{\mu_i}^{\mu_{i+1}} F_1(\tau, y, p) \, d\tau + \sum_{i=1}^{n} \Phi(\mu_i, y_i, \bar{y}_i); \tag{S. 132}$$

$$\dot{y}=g_1(\tau,y,p)\equiv\frac{\varphi_{1\tau}g(\varphi_1,y,1)\,g(\varphi_1,y,p)}{g(\varphi_1,y,1)-g(\varphi_1,y_1,p)},\qquad\text{(S. 133)}$$

where

$$\mu_1=\alpha=\tau(a,a_1),\ \ \mu_n=\beta=\tau(b,b_1);\qquad\text{(S. 134)}$$

$$F_1=F(\varphi_1,y,p)\,\frac{\varphi_{1\tau}g(\varphi_1,y,1)}{g(\varphi_1,y,1)-g(\varphi_1,y,p)}\,;\qquad\text{(S. 135)}$$

$$\Phi(\tau,y,\bar{y})=\int_{\bar{t}}^{t}F(t,\varphi(t,\tau),1)\,dt=\int_{\bar{y}}^{y}\frac{F[\varphi_1(\tau,\xi),\xi,1]}{g(\varphi_1(\tau,\xi),\xi,1)}\,d\xi;\qquad\text{(S. 136)}$$

$$\dot{y}=\frac{dy}{d\tau}\,.\qquad\text{(S. 136a)}$$

Consider the set U^τ of piecewise-smooth lines on which the function $y(\tau)$ is single-valued everywhere, except a finite number of points $\tau=\mu_i$, where $y(\tau)$ may have discontinuities of the first kind. In the (τ,y) plane, the set U^τ is a perfect analog of the set U in the (t,y) plane, as introduced in Chapter II.

We have $U^p\subset U^\tau$. Let $u\in U^\tau$. This line belongs to U^p if

$$\left.\begin{aligned}&p(\tau)\geqslant-1;\\&y(\mu_i+0)-y(\mu_i-0)\geqslant0.\end{aligned}\right\}\qquad\text{(S. 137)}$$

The condition $p(\tau)\leqslant1$ is satisfied automatically, in virtue of the particular choice of the independent variable τ. The vertical segments of the line u in the (τ,y) plane, $\tau=\mu_i$, are the solutions of the equation (S. 107) for $p=1$. It follows from (S. 133) that for $\dot{y}\rightarrow\infty$, we have $p\rightarrow1$. $p(\tau,y,\dot{y})$ can be expressed from (S. 133). Inserting the result in (S. 135), we obtain the function $\widetilde{F}(\tau,y,\dot{y})$. We have

$$W(\tau,y)=\lim_{\dot{y}\rightarrow\infty}\widetilde{F}_1(\tau,y,\dot{y})\,\frac{1}{\dot{y}}=\frac{F(\varphi_1,y,1)}{g(\varphi_1,y,1)}\,.\qquad\text{(S. 138)}$$

The functional $I(u)$, $u\in U^\tau$, is defined by (S. 132). Since the integrand in (S. 136) coincides with the limit (S. 138), and this limit does not depend on the sign of the difference $y_i-\bar{y}_i$ this definition coincides with Definition 1 from § S. 1, and the functional $I(u)$, $u\in U^\tau$, corresponds to type I of the general case, when the function $p\widetilde{F}_1\left(\tau,\,y,\,\dfrac{1}{p}\right)$ exists and is continuous for $p=0$. Therefore all the theorems formulated for this case still apply to this functional.

We introduce the set of (y,z) lines U_0^τ. Here z is the local slope of the line $u_0\in U_0^\tau$ in the coordinates τ,y. To every z corresponds some q defined by the equation

$$z=g_1(\tau,y,q).\qquad\text{(S. 139)}$$

This is a one-to-one correspondence, since $\dfrac{\partial g_1}{\partial q}>0$ for $z=\infty$. Since $g_1(\tau,y,q)$ is continuous, we have $z\rightarrow z_0$ if $q\rightarrow q_0$, and vice versa. The line $u_0\in U_0^\tau$ may therefore be defined either by the pair of functions (y,z), or by the pair (y, q). We have referred to the function q as the local value of the parameter p. By Theorem 1, we may write

$$
\left.
\begin{aligned}
I(u_0) &= \int_\alpha^\beta S(\tau,y,z)\,d\tau + \Phi(\beta,b_1,c) - \Phi(\alpha,a_1,c); \\[2mm]
S &= \widetilde{F}_1(\tau,y,z) - W(\tau,y)z - \int_c^y W_\tau(\tau,\xi)\,d\xi,
\end{aligned}
\right\}
\tag{S.140}
$$

where c is an arbitrary constant, or, using (S. 133) and (S. 135),

$$
I(u_0) = \int_\alpha^\beta S_1(\tau,y,q)\,d\tau + \Phi(\beta,b_1,c) - \Phi(\alpha,a_1,c);
\tag{S.141}
$$

$$
S_1 = \varphi_{1\tau}\,\frac{F(\varphi_1,y,g)\,g(\varphi_1,y,1) - F(\varphi,y,1)\,g(\varphi_1,y,q)}{g(\varphi_1(\tau,y),y,1) - g(\varphi_1(\tau,y),y,q)}
$$

$$
-\int_c^y \varphi_{1\tau}(\tau,\ y)\,\frac{F_t(\varphi_1,y,1)\,g(\varphi_1,y,1) - F(\varphi_1,y,1)\,g_t(\varphi_1,y,1)}{g^2(\varphi_1(\tau,y),y,1)}\,dy.
\tag{S.142}
$$

The theorems of § S. 1 ensure a complete solution of the problem of minimizing I on U^τ. To apply these theorems to the present case, we have to establish a relationship between the sets U_0^τ and $U^p \subset U_0^p$. This relationship is established by the following lemma.

L e m m a 1. Consider a (y, q) line $u_0 \in U_0^\tau$. A necessary and sufficient condition for the existence of a sequence of polygonal lines $\{\gamma_n\} \to u_0$; $\{\gamma_n\} \subset U^p$, i. e., a necessary and sufficient condition for $u_0 \in U_0^p$, is that the functions $q(\tau)$ and $y(\tau)$ satisfy the constraints

$$
q(\tau) \geqslant -1;
\tag{S.143}
$$

$$
\dot{y} - g_1(\tau,\ y,\ q) \geqslant 0
\tag{S.144}
$$

at the points of continuity of $y(\tau)$ and the constraint

$$
y(\mu_i) - \bar{y}(\mu_i) \geqslant 0
\tag{S.145}
$$

at the points of discontinuity μ_i of $y(\tau)$.

N e c e s s i t y. 1) By (S. 137), the sequence $\{\gamma_n\}$ approximating a $(y,\ q)$ line $u \in U_0^\tau$ belongs to U^p if

$$
\left.
\begin{aligned}
q(\tau) &\geqslant -1; \\
y_i - \bar{y}_i &\geqslant 0.
\end{aligned}
\right\}
\tag{S.146}
$$

(S. 143) follows directly from these inequalities.

2) Let at the point $\tau_0 \in (\alpha,\ \beta)$,

$$
\dot{y}(\tau_0) - g_1(\tau_0,\ y(\tau_0),\ q(\tau_0)) < 0.
$$

Let γ_n be an approximating polygonal line sufficiently close to u, and τ_i and τ_{i+1}, $\tau_i \leqslant \tau_0 \leqslant \tau_{i+1}$, are the partition points closest to τ_0. We have

193

$$y_{i+1} - \bar{y}_{i+1} = y_{i+1} - \bar{y}_i - \dot{y}_n(\tau^*) \cdot \Delta\tau_i =$$

$$= [\dot{y}(\tau_0) - g(\tau_0, y(\tau_0), q(\tau_0)) + o(\Delta\tau_i)] \Delta\tau_i. \qquad (\text{S. } 147)$$

Here $\tau^* \in [\tau_i, \tau_{i+1}]$, $\Delta\tau_i = \tau_{i+1} - \tau_i$. Choosing n so that $\Delta\tau_i$ is sufficiently small, we find

$$y_{i+1} - \bar{y}_{i+1} < 0.$$

This signifies that (S. 137) is not satisfied at the point τ_{i+1}, i. e., for $n > N(\Delta\tau_i)$, the sequence $\{\gamma_n\}$ does not belong to U^p. Thus, (S. 143) and (S. 144) are necessary conditions. The necessity of (S. 145) is self-evident.

Sufficiency. Let (S. 143), (S. 144), and (S. 145) apply. The condition $p \leqslant 1$ is satisfied automatically on the entire set U^τ (and therefore on U^p also) because of the particular choice of the independent variable τ. Construct a sequence of polygonal lines $\{\gamma_n\}$ approximating a (y, q) line $u_0 \in U_0^\tau$. We will show that for $n > N$, $\{\gamma_n\} \subset U^p$. To this end it suffices to show that (S. 146) is satisfies for $n > N$. The first condition is true by (S. 143). Now, let the zero closeness function $y(t)$ be continuous and differentiable on (τ_{i-1}, τ_i):

$$y_i - \bar{y}_i = y_{i-1} - \int_{\tau_{i-1}}^{\tau_i} g_1 d\tau = \int_{\tau_{i-1}}^{\tau_i} (\dot{y} - g_1(\tau, y, q)) \, d\tau.$$

By (S. 144)

$$\dot{y} - g_1 \geqslant 0,$$

so that

$$y_i - \bar{y}_i \geqslant 0.$$

Let now $y(\tau)$ have a discontinuity on (τ_{i-1}, τ_i). For sufficiently large n (small $\Delta\tau_{\max}$), this is the only discontinuity on (τ_{i-1}, τ_i). Then

$$y_i - \bar{y}_i = y(\mu) - \bar{y}(\mu) + o(\Delta\tau_i),$$

where μ is the point of discontinuity of $y(\tau)$. By (S. 145), for sufficiently large n,

$$y_i - \bar{y}_i \geqslant 0.$$

Thus, for $n > N$, (S. 146) is satisfied for all $i = 0, 1, \ldots, n-1, n$. Hence $\{\gamma_n\} \subset U^p$. Q. E. D.

A similar lemma for the lines $u \in U_0^T$ is proved by replacing (S. 143), (S. 144), and (S. 145) with the following conditions:

$$q \leqslant 1; \qquad (\text{S. } 148)$$

$$\frac{dy}{dT} - g_2(T, y, q) \leqslant 0 \qquad (\text{S. } 149)$$

at the points where $y(t)$ is continuous, and

$$y(T_i+0)-y(T_i-0)\leqslant 0 \qquad\qquad (\text{S}.150)$$

at the points T_i where $y(T)$ is discontinuous.

Let $\tilde{y}$, $\tilde{q}$ be a pair of functions satisfying the condition

$$S_1(\tau,\tilde{y},\tilde{q})=\inf_{\Gamma_2(\tau)<y<\Gamma_1(\tau);-1<q\leqslant 1,}S(\tau,y,q) \qquad\qquad (\text{S}.151)$$

and y^*, q^* a pair satisfying the condition

$$S_2(T,y^*,q^*)=\inf_{\Gamma_2<y<\Gamma_1;-1<q\leqslant 1.}S_2(T,y,q). \qquad\qquad (\text{S}.152)$$

Here U_0^t, $g_2(T, y, q)$, $S_2(T, y, q)$ are the analogs of U_0^τ, g_1, S_1 in the coordinates T, y.

We now prove the following theorem.

T h e o r e m 2. Let the (y, q) line $\tilde{u}\in U_0^\tau$ defined by the pair $\tilde{y}$, q satisfying (S.151) satisfy conditions (S.143), (S.144), (S.145). Then $\tilde{u}$ is the absolute minimal of the functional (S.106), i. e.,

$$I(\tilde{u})=\inf_{u\in U^p}I(u). \qquad\qquad (\text{S}.153)$$

It follows from expression (S.141) for $I(u)$, $u\in U_0^\tau$, that $\tilde{u}$ is a minimal on the set of the (y, q) lines $u\in U_0^\tau$ satisfying the additional condition $q\geqslant-1$. Since this set encloses U^p, we have $I(\tilde{u})\leqslant I(u), u\in U^p$.

If $\tilde{u}$ satisfies (S.144) and (S.150), there exists a sequence $\{\gamma_n\}\to\tilde{u}$, $\{\gamma_n\}\subset U^p$. Thus, by definition of the lower bound,

$$I(\tilde{u})=\inf_{u\in U^p}I(u).$$

Q. E. D.

A similar proof can be given for Theorem 2*.

T h e o r e m 2*. If a line $u^*\in U_0^T$ defined by the pair y^*, q^* satisfying (S.152) satisfies conditions (S.149), (S.150), we have

$$I(u^*)=\inf_{u\in U^p}I(u). \qquad\qquad (\text{S}.154)$$

C o r o l l a r y. Let the line $u^*\in U_0^t$ defined by (S.152) satisfy (S.148), (S.149), (S.150), i. e., $u^*\in U_0^p$. Then, if $I(\tilde{u})\neq I(u^*)$, where $I(\tilde{u})$ is defined by (S.141), the line $\tilde{u}$ does not satisfy (S.144), (S.145), i. e., $\tilde{u}$ does not belong to U_0^p. The reverse proposition is also true.

Indeed, suppose that these conditions are satisfied on $\tilde{u}\in U_0^\tau$. Then, by Theorems 2 and 2*, $I(u)$, $u\in U^p$, has the lower bounds $I(\tilde{u})$ and $I(u^*)$, in contradiction to the definition of the lower bound.

T h e o r e m 3. Let the functional (S.106) have a minimum on the (y, q) line $\tilde{u}\in U_0^p$. Furthermore, let the following inequalities hold true almost everywhere for the pairs $\tilde{y}$, $\tilde{q}$ and y^*, q^* defined by (S.151) and (S.152):

$$\left.\begin{array}{l}\dfrac{d\tilde{y}}{dt}-g(\tilde{t},\tilde{y},\tilde{q})<0;\\[2ex]\dfrac{dy^*}{dt}-g(t,y^*,q^*)>0.\end{array}\right\} \qquad\qquad (\text{S}.155)$$

The minimal is then a continuous piecewise-smooth function.

Since $\bar{u} \in U_0^p$, we see that $\bar{y}(t)$ and $\bar{q}(t)$ are piecewise-continuous and, therefore, $\bar{u}$ may consist only of a finite number of sections on which one of the following three conditions is satisfied: either 1) $\bar{y}'(t) - g(t, \bar{y}, \bar{q}) > 0$, or 2) $\bar{y}'(t) - g(t, \bar{y}, \bar{q}) = 0$, or 3) $\bar{y}'(t) - g(t, \bar{y}, \bar{q}) < 0$.

By the remark to Theorem 1, condition (S. 151) should be satisfied on those sections where $\bar{y}' - g(t, \bar{y}, \bar{q}) > 0$, i. e., $\bar{y} = \tilde{y}$ and $\bar{q} = \tilde{q}$, but this contradicts (S. 155). Consequently, such sections do not exist. Similar argument proves that sections of type 3 do not exist either. Therefore, we have almost everywhere on $[a, b]$

$$\bar{y}'(t) - g(t, \bar{y}, \bar{q}) = 0, \tag{S. 156}$$

i. e., $\bar{u} \in U^p$.

Corollary. Since the minimal $\bar{u}$ is piecewise-smooth, it can be determined by ordinary classical methods, e. g., the maximum principle. The minimal consists of a finite number of Euler pieces and pieces having the limit direction $p(t) \equiv \pm 1$.

Theorem 4. Let $[\tau_1, \tau_2] \subset [\alpha, \beta]$ be an isolated segment on which the functions $\tilde{y}(\tau)$, $\tilde{q}(\tau)$ obtained from (S. 151) satisfy conditions (S. 144), (S. 145).

We construct the following object $\bar{u}$:

1. On $[a, \tau_1]$, the minimal $\bar{u}$ coincides with the absolute minimal of the functional

$$I_1(u_1, y_1) = \int_a^{\tau_1} F_1(\tau, y, p)\, d\tau - \Phi(\tau_1, y_1), \tag{S. 157}$$

with a free right end $y_1 = y(\tau_1)$.

2. On (τ_1, τ_2), the object $\bar{u}$ is a (y, q) line:

$$\bar{y} = \tilde{y}(\tau); \quad \bar{q} = \tilde{q}(\tau). \tag{S. 158}$$

3. On $[\tau_2, \beta)$, the object $\bar{u}$ coincides with the absolute minimal of the functional

$$I_2(u_2, y_2) = \int_{\tau_2}^{\beta} F_1(\tau, y, p)\, d\tau + \Phi(\tau_2, y_2), \tag{S. 159}$$

with a free left end $y_2 = y(\tau_2)$.

If

$$\left.\begin{array}{l} y_1 \leqslant \tilde{y}(\tau_1); \\[4pt] y_2 \geqslant \tilde{y}(\tau_2), \end{array}\right\} \tag{S. 160}$$

the object $\bar{u}$ is the absolute minimal of the functional (S. 106), i. e.,

$$I(\bar{u}) = \inf_{u \in U^p} I(u). \tag{S. 161}$$

Proof. The functional $I(\bar{u})$ may be written in the form

$$I(\bar{u}) = \int_a^{\tau_1} F_1\, d\tau - \Phi(\tau_1, \bar{y}_1) + \int_{\tau_1}^{\tau_2} S(\tau, \tilde{y}, \tilde{q})\, d\tau +$$

$$+\int_{\tau_2}^{\beta} F_1 d\tau + \Phi(\tau_2,\overline{y}_2) = I_1(\overline{u}_1,\overline{y}_1) + I_2(\overline{u}_2,\overline{y}_2) + \int_{\tau_1}^{\tau_2} S(\tau,\widetilde{y},\widetilde{q})\, d\tau.$$

Similarly, along any line $u \in U^p$,

$$I(u) = I_1(u_1,y_1) + I_2(u_2,y_2) + \int_{\tau_1}^{\tau_2} S(\tau,y,p)\, d\tau.$$

By the conditions of the theorem,

$$I_1(u_1,y_1) - I_1(\overline{u}_1,\overline{y}_1) \geqslant 0;$$

$$I_2(u_2,y_2) - I_2(\overline{u}_2,\overline{y}_2) \geqslant 0;$$

$$\int_{\tau_1}^{\tau_2} S(\tau,y,p)\, d\tau - \int_{\tau_1}^{\tau_2} S(\tau,\widetilde{y},q)\, d\tau = \int_{\tau_1}^{\tau_2} [S(\tau,y,p) - S(\tau,\widetilde{y},\widetilde{q})]\, d\tau \geqslant 0,$$

and therefore

$$I(u) - I(\overline{u}) \geqslant 0.$$

On the other hand, by the conditions of the theorem and by Lemma 1, we see that there exists a sequence $\{\gamma_n\} \subset U^p$ such that $I(\gamma_n) \underset{n \to \infty}{\longrightarrow} I(\overline{u})$. Therefore, in virtue of the definition of the lower bound, (S. 161) applies. Q. E. D.

Remark. If $t_1 = a$ or $t_2 = b$, (S. 160) is replaced only by the first or the second inequality, respectively.

Corollary. If (S. 160) is now satisfied at the points τ_1 and τ_2, the extremal pair $(\overline{y}(\tau),\ \overline{q}(\tau))$ does not necessarily coincide with $\widetilde{y}(\tau),\ \widetilde{q}(\tau)$ on $[\tau_1, \tau_2]$, but some of its properties emerge from Theorem 3.

1. If there exist at least two points $\tau = \xi_1$, $\tau = \xi_2$ where

$$y(\tau) = \widetilde{y}(\tau), \tag{S. 162}$$

for $\tau \in [\xi_1, \xi_2]$ we have $\overline{y} = \widetilde{y}$ and $\overline{q} = \widetilde{q}$.

2. If condition (S. 160) is not satisfied at a single point, τ_1 say, we have $\overline{y}(\tau) \geqslant \widetilde{y}(\tau)$ for $\tau \in [\tau_1, \tau_2]$. If at least one root $\xi \in [\tau_1, \tau_2]$ of equation (S. 162) exists in this case, for $\xi \leqslant \tau \leqslant \tau_2$ we have $\overline{y} = \widetilde{y}$, $\overline{q} = \widetilde{q}$. An analogous theorem clearly can be proved for an isolated segment $[t_1, t_2]$ on which the functions $\widetilde{y}(t)$ and $\widetilde{q}(t)$ defined by (S. 152) satisfy (S. 149) and (S. 150). Inequalities (S. 160) should reverse their sign in this case.

Discussion. 1. It follows from the theory that the functional (S. 106) on the set U^p of functions with a bounded derivative has $(y,\ q)$ minimals with a "branched" derivative, similar to the $(y,\ z)$ minimals of the simplest functional. The role of the vertical directions in this case is assumed by the directions $p = \pm 1$.

2. As for the simplest functionals of type I (the function $pF\left(t,\ y,\ \dfrac{1}{p}\right)$ is continuous for $p = 0$), the solution of the variational problem should not start with the solution of Euler's equations, i. e., we should not attempt to find a weak local minimal on the class C_1. A better approach is to set up the function $S_1(t,\ y,\ q)$ (or $S_2(t,\ y,\ q)$) and to find its minima $\widetilde{y}(\tau),\ \widetilde{q}(\tau)$ for every fixed $\tau \in [\alpha, \beta]$. If the functions $\widetilde{y}(\tau),\ \widetilde{q}(\tau)$ satisfy the conditions of Theorem 2

(Theorem 2*), this completes the solution. The functions $\tilde{y}(\tau)$, $\tilde{q}(\tau)$ define a (y, q) line $\tilde{u} \in U_0^p$ on which the function (S. 106) has an absolute minimum. If almost everywhere on $[\alpha, \beta]$,

$$\dot{\tilde{y}} - g_1(\tau_1, \tilde{y}, \tilde{q}) = 0,$$

we have $\tilde{u} \in U_0^p$. Otherwise, the pair $\tilde{y}$, $\tilde{q}$ defines a minimizing sequence $\{\gamma_n\} \in U^p$. For any given $\varepsilon > 0$, there exists N such that for $n > N$,

$$|I(u) - I(\tilde{u})| < \varepsilon,$$

where u is any line from U^p.

3. If the conditions of Theorem 2 are not satisfied on $\tilde{u}$ and u^*, Theorem 4 enables us to identify pieces of the absolute minimal on those segments where the corresponding conditions hold true. To obtain a complete solution of the problem in this case, we have to find the pieces of minimal on those sections where the conditions of Theorem 4 do not hold true. These pieces coincide with the minimals of the corresponding functionals (S. 157) and (S. 159) defined on the corresponding sections.

4. If the lines $\tilde{u}$ and u^* satisfy the conditions of Theorem 3, the minimal $\bar{u}$ of the functional (S. 106) is piecewise-smooth when $\tilde{u} \in U_0^p$ and it can be found by conventional classical methods, e. g., by Pontryagin's maximum principle.

5. If we compare the results with those of § S. 1, we see that the functionals (S. 106) with regard to their extremal properties are the closest to type I functionals of the general case.

The strongest direct expression of this analogy is provided by Theorem 2. However, unlike the type I functionals, functional (S. 106) may also have smooth Euler minimals, if the conditions of Theorem 2 (Theorem 2*) are not satisfied. In this respect, functional (S. 106) is closer to the simplest functional of type II of the general case ($f(t, y, p, 1)$ has a discontinuity of the first kind for $p = 0$). There is a highly significant difference between the two types: for type II functionals the minimum of S along the (y, z) line u_0 is only a necessary condition of a minimum of $I(u)$ on the line u_0, and in the presence of this minimal the absolute minimum may be attained on an ordinary Euler extremal, whereas for the other type, as for simplest functionals of type I, the minimum of S on a (y, q) line $u \in U_0^p$ is a sufficient condition for the absolute minimum of the functional on this line.

§ S. 3. OPTIMAL PROGRAM FOR HORIZONTAL FLIGHT OF AN AIRCRAFT

We will now consider the optimal thrust control for the horizontal flight of a jet aircraft over a maximum range. This problem was dealt with by Hibbs /15/ and by Miele and Cicala /14/. According to their results, assuming a linear dependence of thrust on fuel consumption,

$$P = j\beta, \tag{S. 163}$$

where P is the thrust, β is the fuel consumption, $j=$ const is the effective nozzle velocity, we can reduce the original problem to extremizing a functional of the form

$$\int_{m_0}^{m_{\text{K}}} [A(V,\,m)+V'B(V,\,m)]\,dm, \qquad (\text{S.}\,164)$$

where m is the instantaneous aircraft mass, V is the aircraft velocity, $V'=\dfrac{dV}{dm}$. Euler's equation for this functional degenerates to a finite equation

$$\frac{\partial A}{\partial V} - \frac{\partial B}{\partial m}=0, \qquad (\text{S.}\,165)$$

which in the general case does not pass through the given initial and terminal points $(V_0,\,m_0)$, $(V_t,\,m_t)$ in the $(m,\,V)$ plane.

An ingenious application of Green's theorem enabled Miele to construct the sought solution and to show that it is made up of pieces satisfying equation (S. 165) or the conditions $\beta=\beta_{\max}$ and $\beta=0$. For the case of a non-linear dependence $P=P(\beta)$ the solution is no longer degenerate and the method of Lagrange's multipliers must be applied.

In the present section, this problem is solved by a different mathematical approach, namely by the theory presented in § S. 1 and § S. 2.

A quite general dependence $P=P(\beta)$ is assumed. As in /15/, we seek a dependence $V=V(m)$ ensuring the maximum range for given values of m_0, V_0 and m_t, V_t. It is shown that this problem of maximizing a functional can be reduced to maximization of a function of two variables $S(m,\,V,\,\beta)$ for every fixed $m \in (m_0,\,m_t)$, where V and β are assumed independent. The optimal program obtained in this way is found to be degenerate, i. e., independent of the position of the end points $(m_0,\,V_0)$ and $(m_t,\,V_t)$. In case of a linear characteristic $P=j\beta$, $j=$ const, S is independent of β and the solution $V=\overline{V}(m)$ coincides with that from /14, 15/. If $P=P(\beta)$ is nonlinear, the absolute maximum range is attained with the "pulsed thrust" program, which in fact constitutes the (y,q) extremal that we mentioned before. This program amounts to the following: the aircraft starts from the initial state $(m_0,\,V_0)$ with its engine off $(\beta=0)$ or alternatively with maximum thrust $\beta=\beta_{\max}$ until it reaches some curve $V=\overline{V}^0(m)$ in the $(m,\,V)$ plane. After that, the optimal program reduces to an alternating succession of powered sections with some optimal thrust $P(\overline{\beta})=$ const and coasting sections with cut engines $(\beta=0,\,m=$ const$)$. The engine is switched on whenever the coasting velocity has dropped to $V=\overline{V}^0(m)$. The engine switching frequency should be as high as possible. The higher the switching frequency, the deeper is the maximum. This program is continued until the aircraft reaches the line $\beta=0$ $(m=$ const$)$ or $\beta=\beta_{\max}$ passing through the point $(m_t,\,V_t)$ for $\overline{V}^0(m_t)>V_t$ or $V_0(m_t)<V_t$, respectively.

3. 1. Statement of the problem

The equations of motion of a jet aircraft in horizontal rectilinear motion are

$$m\dot{V} - P(h,\ V,\ \beta) + X(h,\ V,\ Y) = 0;$$
$$Y - mg_{\tau} = 0;$$
$$\dot{m} = -\beta;$$
$$\dot{x} = V,$$
$$(\text{S. } 166)$$

where m is the aircraft mass, g_{τ} is the gravitational acceleration, x is the horizontal coordinate of the aircraft, V is the aircraft velocity, h is the altitude of horizontal flight, X is the drag, Y is the lift, $\dot{V} = \dfrac{dV}{dt}$ and $\dot{m} = \dfrac{dm}{dt}$.

We adopt the following hypotheses regarding the forces entering (S. 166):

1. The aerodynamic forces X and Y are independent of the aircraft acceleration (the aerodynamic lag is ignored). Since the altitude is constant, X and Y are functions of velocity only. X is furthermore a function of the lift Y.

2. The thrust P is a function of the velocity V and the per-second fuel consumption β. It may be written in the form

$$P(V,\ \beta) = f_1(V) f_2(\beta),$$

where $f_1(V)$ is the velocity characteristic of thrust — an arbitrary positive function; $f_2(\beta)$ is the fuel consumption characteristic of thrust — an increasing function, generally displaying the property $f_2(\beta)\big|_{\beta=0} \leqslant 0$.

The physical meaning of $P(V, 0)$ is back-pressure. The dependence $P(h)$ is of no consequence, since

$$h = \text{const.}$$

3. The gravitational acceleration g_{τ} is constant. Equations (S. 166) are in fact written assuming constant g and ignoring the effects of the Earth's spin. The aircraft in these equations is regarded as a point mass.

4. The fuel consumption β may vary between 0 and $\beta_{\max}$.

Let us formulate the boundary conditions. We assume that at the initial time $t=0$ the aircraft mass is m_0 and the velocity is V_0. The point of origin is chosen so that $x(0) = 0$.

At the end of the flight, $t = t_t$, we have

$$m = m_t,\ V = V_t,\ x = x_t,$$

where neither t_t nor x_t are fixed. Our problem is to find a system of functions $x(t)$, $V(t)$, $m(t)$, $\beta(t)$, $Y(t)$ satisfying equations (S. 116) which ensure a maximum range x_t on the set of pentades

$$[x(t),\ V(t),\ m(t),\ \beta(t),\ Y(t)]$$

satisfying (S. 166).   We are dealing with five unknown functions and four equations, i. e., the system has one degree of freedom.   The last equation in (S. 166) gives an expression for the functional to be maximized

$$x_t=\int\limits_{0}^{t_t} V\,dt= -\int\limits_{t_t}^{0} V\,dt=\int\limits_{m_t}^{m_0} \frac{V}{\beta}\,dm. \tag{S. 167}$$

Eliminating Y and dt between the first three equations in (S. 166), we find

$$\frac{dV}{dm}=-\frac{1}{m\beta}\,[P(V,\ \beta)-X(V,\ m)]. \tag{S. 168}$$

The problem thus reduces to maximizing the functional

$$I=\int\limits_{m_t}^{m_0} \frac{V}{\beta}\,dm \tag{S. 167*}$$

under the constraints

$$V'\equiv\frac{dV}{dm}=-\frac{1}{m\beta}\,[P(V,\ \beta)-X(m,\ V)]; \tag{S. 168*}$$

$$0\leqslant\beta\leqslant\beta_{\max}; \tag{S. 169}$$

$$V(m_t)=V_t;\quad V(m_0)=V_0. \tag{S. 170}$$

Problems of this kind were previously considered in § S. 2.   Here we have

$$F(m,\ V,\ \beta)=\frac{V}{\beta}\ ; \tag{S. 171}$$

$$g(m,\ V,\ \beta)=-\frac{1}{m\beta}\,[P(V,\ \beta)-X(m,\ V)]. \tag{S. 172}$$

In virtue of the above properties of the function $P(V,\ \beta)$,

$$\frac{\partial g}{\partial\beta}<0 \tag{S. 173}$$

for all $\beta\in[0,\ \beta_{\max}]$.

To finally reduce this problem to the form discussed in § S. 2, it suffices to replace β with an auxiliary parameter

$$p=2\,\frac{\beta}{\beta_{\max}}-1\,. \tag{S. 174}$$

Inequalities (S. 169) then take the form

$$|p|\leqslant1, \tag{S. 175}$$

and condition (S. 173) reduces to

$$\frac{\partial g}{\partial p}>0. \tag{S. 176}$$

For $p=1$, i. e., $\beta=\beta_{max}$, the derivative $V'=g(m,\ V,\ \beta)$ takes on the least admissible value at the point $(m,\ V)$ of the phase plane, and for $p=1$, $\beta=0$ it takes the largest value:

$$V'=g(m,\ V,\ \beta)\to\infty\ \text{for}\ \beta\to0. \tag{S. 177}$$

Physically (S. 177) signifies the fact that the limit value $\beta=0$ corresponds to flight with constant mass ($m=$ const), i. e., motion of the representing point along the vertical in the phase plane ($m,\ V$). The latter factor signifies that the independent variable m has the properties of the auxiliary variable τ introduced in § S. 3 (constancy on the limit direction $\beta=0$).

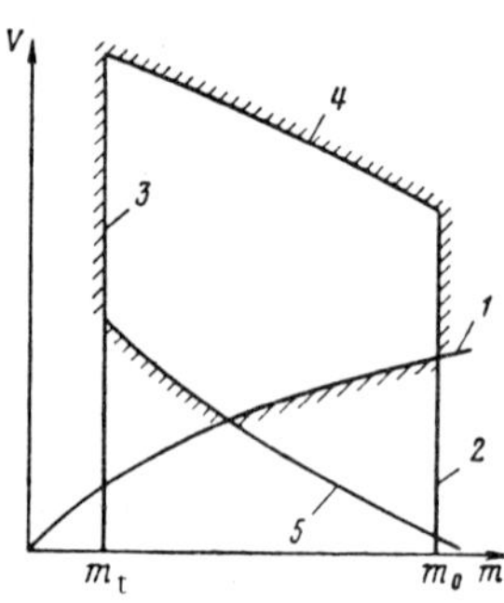

FIGURE S. 3

Let us now establish the boundary of the region B of the physically admissible values of the function $V(m)$. The boundary is made up of pieces of the following lines (Figure S. 3):

Line 1: $mg_\tau-Y(V,\ \alpha_{max})=0$, where α_{max} is the maximum admissible angle of attack. This line is the lower limit for the admissible values of the function $V(m)$.

Lines 2 and 3: the vertical segments $m=m_0$ and $m=m_t$.

Lines 4 and 5: lines of flight with maximum thrust $\beta=\beta_{max}$ passing through the points $(m_0,\ V_0)$ and $(m_t,\ V_t)$ which give, respectively, the upper ,and the lower limit of the admissible values of $V(m)$.

3. 2. Optimal control program

Let us first determine the function $S(m,\ V,\ q)$. Here $V(m)$ is the zero closeness function of the sough extremal, $q(m)$ is the local value of the fuel consumption β. Since the independent variable m is also the parameter, we may use expression (S. 49) for S. Thus,

$$S=F-Wg(m,\ V,\ \beta)-\int_c^V W_m(m,\ \xi)\,d\xi; \tag{S. 178}$$

$$W=\frac{F(m,V,0)}{g(m,V,0)}. \tag{S. 179}$$

Using (S. 171) and (S. 172), we may write

$$W(m,\ V)=\frac{mV}{\bar{x}(m,V)}; \tag{S. 180}$$

$$\bar{x}(m,\ V)=X-P(V,\ 0); \tag{S. 181}$$

$$S(m,\ V,\ q)=\frac{V}{q}+\frac{mV}{\bar{x}}\ \frac{1}{mq}[P(V,\ q)-X(m,\ V)]-$$

$$-\int_c^V W_m(m,\ \xi)\,d\xi,$$

or, using (S. 181),

$$S = \frac{V f_1(V) j(q)}{\bar{x}(m, V)} - \int_c^V \left[\frac{m\xi}{X(m, \xi) - P(\xi, 0)} \right]_m d\xi, \qquad \text{(S. 182)}$$

where

$$j(\beta) = \frac{f_2(\beta) - f_2(0)}{\beta}. \qquad \text{(S. 183)}$$

For an ideal liquid-propellant engine, j is independent of β and gives the ideal nozzle velocity.

Let us find the pair of functions $\bar{V}(m)$, $\bar{q}(m)$ on which $S(m, V, q)$ attains its absolute minimum over the set of admissible values of V and q for every fixed $m \in [m_t, m_0]$.

Writing the thrust in the form

$$P(V, \beta) = f_1(V) f_2(\beta),$$

we can devise an attractively simple method of solution of the last problem. In fact, the point $q = \bar{q}$ on which S attains its maximum for any m, $V \in B$ is independent of m and V coincides with the maximum of the function $j(\beta)$ on $0 \leqslant \beta \leqslant \beta_{max}$. Figures 4a and 4b show two alternative fuel consumption characteristics of the thrust program. Draw a family of imaginary rays from the point $(0, f_2(0))$ which have at least one common point with the curve of $f_2(\beta)$ on the semi-interval $(0, \beta_{max})$. In this family, we select a ray γ which makes the largest slope angle to the horizontal axis. The abscissas $\bar{q}_1, \ldots, \bar{q}_t$ of the intersection points of this ray with the curve $f_2(\beta)$ on the semi-interval $(0, \beta_{max})$ give the maximum of $j(\beta)$, i. e., S, as we see from (S. 183). Thus, if $P(V, \beta)$ is representable in the form of a product $f_1(V) f_2(\beta)$, the optimal value $\beta = \bar{\beta}$ is independent of m and V.

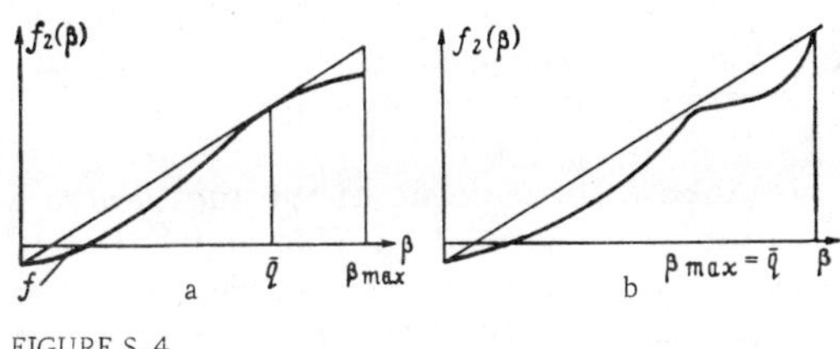

FIGURE S. 4

The optimal dependence $V^0 = \tilde{V}^0(m)$ is specified by the condition of absolute maximum of the function of single variable $S(m, j_{max}, V^0)$ in B for every fixed m. As a necessary condition, it should consist of pieces of the boundary of B and of continuous pieces satisfying the finite equation.

$$S_V = \left[\frac{f_1(V) j_{max}}{X(V, m) - P(V, 0)} \right]_V - \left[\frac{mV}{X(V, m) - P(V, 0)} \right]_m = 0 \qquad \text{(S. 184)}$$

which are joined by vertical segments at the points $m=\mu_i$ $(i=1,\,2,\,\ldots,\,r)$ where

$$S(\mu_i,\ j_{\max},\ V_1(\mu_i))=S(\mu_i,\ j_{\max},\ V_2(\mu_i)), \tag{S.185}$$

$V_1(m)$ and $V_2(m)$ being two solutions of (S.184).

Equation (S.184) coincides with the equation $W=0$, the so-called "singular curve" from /14/, i.e., it coincides with Euler's degenerate equation for the case of linear dependence of thrust on fuel consumption, if we take $P(V,\,0)=0$, $f_1(V)\equiv1$ and $j_{\max}=V_2=$ const, the effective nozzle velocity, is assumed constant. Here the first two conditions are the simplifying assumptions adopted in /14/. Consequently, we can apply the results of /14/ for qualitative estimate of the zero closeness function and the optimal $V=\bar{V}^0(m)$. In other words, all the properties of the extremal $\bar{V}(m)$ investigated in detail in /14/ are applicable:

1) Equation (S.184) has two solutions: supersonic $V^0=V_1^0(m)>a$ and subsonic $V^0=V_2^0(m)<a$, where a is the velocity of sound at altitude h. Along both solutions $\dfrac{dV}{dm}>0$ (motion with decreasing velocity), and $V_2^0(m)$ invariably passes through the origin.

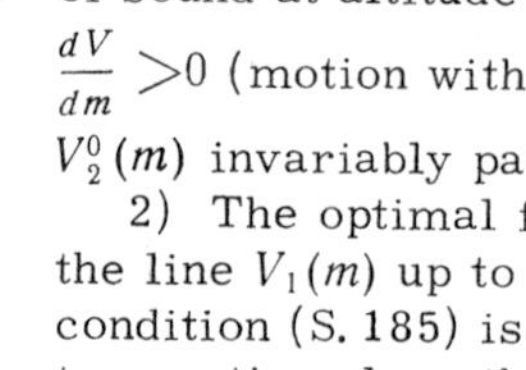

FIGURE S.5

2) The optimal flight program first follows the line $V_1(m)$ up to the point $m=\mu$, where condition (S.185) is satisfied. Then it changes to coasting along the vertical $m=\mu$ until it reaches the line $V_2(m)$, and then it proceeds along the line $V_2(m)$.

The motion from the point $(m_0,\,V_0)$ to the solution of (S.184) and from the latter to the point $(m_t,\,V_t)$ proceeds along the appropriate piece of the boundary of B, i.e., either with $\beta=0$ or with $\beta=\beta_{\max}$, according as the points $(m_0,\,V_0)$ and $(m_t,\,V_t)$ are respectively located below or above the line $V^0=\bar{V}^0(m)$ (Figure S.5).

According to our results, the entire optimal function $V=\bar{V}^0(m)$ is determind (with necessity and sufficiency) by the condition of the absolute maximum of the function of a single variable $S(m,\,j_{\max},\,V)$ in B for every fixed m.

We have thus obtained a local value of the fuel consumption $\tilde{q}$ on the inclined sections, which is independent of m and V and corresponds to the maximum value of j. We have also obtained the function $\bar{V}^0(m)$ which coincides with the optimal velocity $\bar{V}(m)$ if constant nozzle velocity $j_{\max}$ is assumed.

By Theorem 2, the pair $\tilde{V}(m)$, $\tilde{q}(m)$ defines the absolute $(y,\,q)$ minimal $\bar{u}\in U_0$ if it satisfies the inequalities

$$\frac{d\tilde{V}}{dm}-g(m,\,\tilde{V},\,\tilde{q})\geqslant0; \tag{S.186}$$

$$\tilde{V}(\mu+0)-\tilde{V}(\mu-0)\geqslant0, \tag{S.187}$$

where μ is the point of discontinuity of $\tilde{V}(m)$. Inequality (S.186) is always satisfied, since the results of /14/ show that if the minimal $\tilde{V}(m)$ contains pieces of both branches, $V_1(m)$ and $V_2(m)$, motion always proceeds first along the supersonic branch $V_1(m)$ and then along the subsonic branch $V_2(m)$. It thus remains to check (S. 187).

As we have mentioned before, $\frac{dV^0}{dm}>0$ along both extremal branches. On the other hand, according to (S. 172), $g\leqslant0$ whenever

$$P(V,\ \beta)\geqslant X. \qquad\qquad (S.\ 188)$$

The last inequality always holds true, i. e., the maximum j as a rule corresponds to a thrust which is greater than the drag. In this case, (S. 186) is naturally satisfied. The drag X increases with increasing V and m, so that (S. 188) may fail, if at all, on the supersonic branch $V_1(m)$. But on this branch $\frac{dV^0}{dm}$ also reaches its maximum value (see /14/), and (S. 186) is satisfied even if (S. 188) does not apply.

The pair $\widehat{V}(m)$, $\tilde{q}(m)$ thus constitutes the sought solution of the problem. It defines a $(V,\ q)$ line $\overline{u}\subset U_0$ on which the range has its absolute maximum.

Let us examine the physical meaning of this $(V,\ q)$ maximal $\overline{u}$.

Each approximating polygonal line $\gamma_n\subset U$ represents motion with periodic thrust switching. The inclined sections of the polygonal line γ_n correspond to powered flight with fuel consumption $\beta=\tilde{q}(m_i)$, and the vertical sections represent coasting with the engine cut.

The engine is always switched on as soon as the coasting velocity with $m=m_i$ reaches the value $V^0(m_i)$, and it is cut off when the mass drops to m_{i+1}. The engine is switched on a total of n times.

We refer to this program as "pulsed thrust" program with frequency n, mean velocity $V^0(m)$, and power thrust q. The $(V,\ q)$ line $u\subset U_0$ represents a pulsed thrust program of infinite frequency.

Let us now consider the extremal for various typical particular cases.

I. There is a finite number of values $\overline{q}_i$ $(i=1,2,\ldots,k)$ satisfying the condition $j(\beta)=j_{max}$ (a finite number of intersection points between the ray γ and the characteristic $f_2(\beta)$ on $(0,\ \beta_{max})$). All $\overline{q}_i$ satisfy inequality (S. 186) everywhere on $[m_t,\ m_0]$.

The function $\beta^0(m)$ defined by the equation

$$\frac{d\widetilde{V^0}(m)}{dm}=-\frac{1}{m\beta^0}\ [P(\widetilde{V}^0,\ \beta)-X(\widetilde{V}^0,\ m)] \qquad\qquad (S.\ 189)$$

continuously varies with m. Since, on the other hand, any of the optimal values $q=\overline{\beta}_i=\text{const}$, we have $g(m,\ V,\ q)\neq\frac{d\widetilde{V^0}}{dm}$. Hence, the optimal program is a $(\widehat{V},\ q)$ line in the $(m,\ V)$ plane, i. e., the "pulsed thrust" program of infinite frequency. The representing point in the $(m,\ V)$ plane should therefore move from position $(m_0,\ V_0)$ along the boundary of B ($\beta=0$ or $\beta=\beta_{max}$) until it reaches the zero closeness line $V=\widehat{V}^0(m)$.

After that, the "pulsed thrust" program begins with maximum permissible frequency. The fuel consumption on the powered sections should be equal to any one of the $\overline{q}_i$ values, and the mean velocity should coincide with the optimal function $V=\overline{V}^0(m)$. This program should be continued until the representing point again reaches the boundary. Then the motion continues along the boundary until the terminal position $(m_t,\ V_t)$ is reached.

205

II.   There is a finite number of values $\bar{q}_i$, $i=1, 2, \ldots, k$.

Some of them satisfy inequality (S. 186), whereas others do not satisfy this inequality. The sought optimal is a "pulsed thrust" program, similar to that described under I. The fuel consumption on the powered sections should be equal to any of the $\bar{\beta}_i$ satisfying (S. 184).

III.   There is a continuum of values

$$0 \leqslant \bar{\beta}_0 \leqslant \bar{\beta}_t \leqslant \beta_{max}$$

for which $j(q) = j_{max}$ (the ray γ has a common segment with the thrust characteristic $f_2(\beta)$, Figure S. 6). Part of the segment $[\bar{\beta}_0, \bar{\beta}_t]$ satisfies inequality (S. 188), where the other part does not necessarily satisfy this inequality.

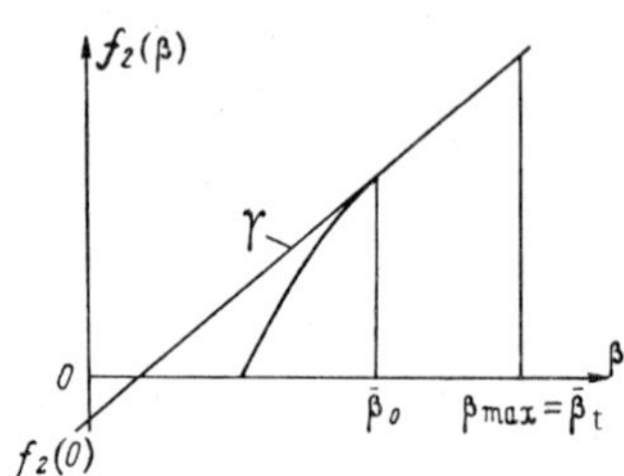

FIGURE S. 6

The particular case $\bar{\beta}_0 = 0$, $\bar{\beta}_t = \beta_{max}$ here corresponds to a linear thrust characteristic. In this case, infinitely many optimal solutions exist. They correspond to "pulsed thrust" programs with infinite switching frequency and a common zero closeness line $V = V^0(j_{max}, m)$, which are analogous to those described under I, but differing in that they may have any dependence $q(m) \in [\bar{\beta}_0, \bar{\beta}_t]$ as long as it satisfies inequality (S. 188). As a particular case, we may even take a dependence $q = \beta^0(m)$ which converts the inequality in (S. 188) into equality. The "pulsed thrust" program then degenerates into Hibbs's continuous control /15/ originally derived for a linear characteristic $P(\beta)$, and the optimal function $\bar{V}(m)$ coincides with the zero closeness line $V^0(m)$.

The physical meaning of the multiple solutions in this case is that the range is affected only by the mean velocity $V^0(m)$, and is insensitive to local deviations from the mean.

However, as the switching frequency is of necessity finite in practice, a single minimal should be selected among the infinitely many optimals, on which the range attains a strict maximum. This minimal coincides with Hibbs's continuous thrust control characteristic.

In mathematical terms, the infinity of solutions implies that $j = $ const on $[\bar{\beta}_0, \bar{\beta}_t]$ and therefore the function $S(m, j, V)$ is independent of q, being a function of a single variable V for every fixed m.

Bibliography for Supplement

1. Bogolyubov, N.N. Sur quelques methodes nouvelles dans le calcul des variations. — Annali di matem. 1930.
2. Ermilin, K.S. K voprosy o razryvnykh resheniyakh v variatsionnom ischislenii (Discontinuous Solutions in Variational Calculus). — Trudy Tbilisskogo Matematicheskogo Instituta, Vol. I. 1937.
3. Ermilin, K.S. Ob odnoi zadache variatsionnogo ischisleniya (A Certain Problem of Variational Calculus). — Trudy Tbilisskogo Matematicheskogo Instituta, Vol. IV. 1938.
4. Ermilin, K.S. O nekotorykh zadachakh variatsionnogo ischisleniya (Some Problems of Variational Calculus). — Uchenye Zapiski LGU, No. 111., math. ser., Issue No. 16. 1949.
5. Kerimov, M.K. O dvumernykh razryvnykh zadachakh variatsionnogo ischisleniya (Two-Dimensional Discontinuous Problems of Variational Calculus). — Trudy Matematicheskogo Instituta AN GrSSR, No. 18. 1951.
6. Kerimov, M.K. O neobkhodimykh usloviyakh ekstremuma v razryvnykh variatsionnykh zadachakh s podvishnymi kontsami (Necessary Extremum Conditions in Discontinuous Variational Problems with Free Ends). — Doklady AN SSSR, 79. 1951.
7. Kerimov, M.K. K teorii razryvnykh variatsionnykh zadach s podvizhnymi kontsami (Theory of Discontinuous Variational Problems with Free Ends). — Doklady AN SSSR, 136, No. 3. 1961.
8. Courant, R. and D. Hilbert. Methods of Mathematical Physics. New York, Interscience, Vol. 1. 1953; Vol. 2. 1962.
9. Krylov, N.M. and Ya. Tamarkin. Sur le methode de W. Ritz. — Izvestiya Rossiiskoi Akademii Nauk, Nos. 2 — 3. 1918.
10. Lavrent'ev, M.A. Sur quelques problemes du calcul des variations. — Annali di matem. 1926.
11. Lyusternik, L.A. Über einige Anwendungen d. direkten Methoden in Variations rechtung. — Matematicheskii Sbornik, XXXIII. 1926.
12. Nikoladze, G.N. Einige Eigenschaften der Risspunktkurve in der Variationsrechnung. — Tbilissi Bulletin.
13. Razmadze, A.M. Sur les solutions discontinues dans le calcul des variations. — Math. Ann., 94. 1925.
14. Cicala, P. and A. Miele. Generalized Theory of the Optimum Thrust Programming for the Lever Flight of a Rocket-Powered Aircraft. — Jet Propulsion, 26, No. 6. 1956.
15. Hibbs, A. Optimum Burning Program for Horizontal Flight. — Journal of American Rocket Soc., 22, No. 4. 1952.
16. Ritz. Uber eine neue Methode zur Losung gewisser Variationsprobleme der mathematischen Physik. — Journal fur die reine und angew. Mathem., 135. 1909.
17. Tonelli, L. Fondamenti di calcolo delle variazioni, Bologna, 1 — 2, No. 7. 1921 — 1923.

REFERENCES USED IN THE PREPARATION OF
THE BOOK

Krotov, V.F. Metody resheniya variatsionnykh zadach na osnove
    dostatochnykh uslovii absolyutnogo minimuma (Methods of
    Solution of Variational Problems from the Sufficient Conditions
    of the Absolute Minimum), Part I. — Avtomatika i Telemekhanika,
    No.12. 1962; Part II — Avtomatika i Telemekhanika, No.5. 1963;
    Part III — Avtomatika i Telemekhanika, No.7. 1964.
Krotov, V.F. Nekotorye novye metody variatsionnogo ischisleniya
    i ikh prilozheniya k dinamike poleta (Some New Variational
    Methods and Their Application to Flight Dynamics). — Dissertation.
    1963.
Krotov, V.F. Priblizhennyi metod sinteza optimal'nogo upravleniya
    (An Approximate Method of Optimal Control Synthesis). — Avtomatika
    i Telemekhanika, No.11. 1964.
Krotov, V.F. Osnovnaya zadacha variatsionnogo ischisleniya dlya
    prosteishego funktsionala na sovokupnosti razryvnykh funktsii
    (The Fundamental Problems of Variational Calculus for the
    Simplest Functional on the Set of Discontinuous Functions). —
    Doklady AN SSSR, 137, No.1. 1961.
Krotov, V.F. Novyi metod variatsionnogo ischisleniya i nekotorye
    ego prilozheniya (A New Method of Variational Calculus and Some
    of Its Applications). — Dissertation. 1962.
Krotov, V.F. Ob optimal'nom rezhime gorizontal'nogo poleta samoleta
    (Optimal Horizontal Flight of an Aircraft). — In: "Mekhanika."
    Oborongiz. 1961.
Krotov, V.F. Dostatochnye uslviya optimal'nosti dlya diskretnykh
    upravlyaemykh sistem (Sufficient Conditions of Optimality for
    Discrete Control Systems). — Doklady AN SSSR, 172, No.1. 1967.
Bukreev, V.S. Ob odnom metode priblizhennogo sinteza optimal'nogo
    upravleniya (A Method of Approximate Synthesis of Optimal
    Control). — Avtomatika i Telemekhanika, No.11. 1968.
Gurman, V.I. Ob optimal'nykh protsessakh osobogo upravleniya
    (Optimal Processes of Singular Control). — Avtomatika i Tele-
    mekhanika, No.5. 1965.
Gurman, V.I. Metod issledovaniya odnogo klassa optimal'nykh
    skol'zyashchikh rezhimov (Investigation of a Certain Class of
    Optimal Sliding Controls). — Avtomatika i Telemekhanika, No.7.
    1965.
Gurman, V.I. Metod kratnykh maksimumov i usloviya otnositel'noi
    optimal'nosti vyrozhdennykh rezhimov (The Method of Multiple
    Maxima and the Conditions of the Relative Optimum of Degenerate
    Controls). — Avtomatika i Telemekhanika, No.12. 1967.